PREPARING FOR THE HIGH FRONTIER
THE ROLE AND TRAINING OF NASA ASTRONAUTS IN THE POST-SPACE SHUTTLE ERA

Committee on Human Spaceflight Crew Operations

Aeronautics and Space Engineering Board

Division on Engineering and Physical Sciences

NATIONAL RESEARCH COUNCIL
OF THE NATIONAL ACADEMIES

THE NATIONAL ACADEMIES PRESS
Washington, D.C.
www.nap.edu

THE NATIONAL ACADEMIES PRESS **500 Fifth Street, N.W.** **Washington, DC 20001**

NOTICE: The project that is the subject of this report was approved by the Governing Board of the National Research Council, whose members are drawn from the councils of the National Academy of Sciences, the National Academy of Engineering, and the Institute of Medicine. The members of the committee responsible for the report were chosen for their special competences and with regard for appropriate balance.

This study is based on work supported by Contract NNH10CC48B between the National Academy of Sciences and the National Aeronautics and Space Administration. Any opinions, findings, conclusions, or recommendations expressed in this publication are those of the authors and do not necessarily reflect the views of the agency that provided support for the project.

Cover: Design by Tim Warchocki. Images courtesy of the National Aeronautics and Space Administration.

International Standard Book Number-13: 978-0-309-21869-6
International Standard Book Number-10: 0-309-21869-1

Copies of this report are available free of charge from:

Aeronautics and Space Engineering Board
National Research Council
500 Fifth Street, NW
Washington, DC 20001

Additional copies of this report are available from the National Academies Press, 500 Fifth Street, NW, Lockbox 285, Washington, DC 20055; (800) 624-6242 or (202) 334-3313 (in the Washington metropolitan area); Internet, http://www.nap.edu.

Printed in the United States of America.

THE NATIONAL ACADEMIES

Advisers to the Nation on Science, Engineering, and Medicine

The **National Academy of Sciences** is a private, nonprofit, self-perpetuating society of distinguished scholars engaged in scientific and engineering research, dedicated to the furtherance of science and technology and to their use for the general welfare. Upon the authority of the charter granted to it by the Congress in 1863, the Academy has a mandate that requires it to advise the federal government on scientific and technical matters. Dr. Ralph J. Cicerone is president of the National Academy of Sciences.

The **National Academy of Engineering** was established in 1964, under the charter of the National Academy of Sciences, as a parallel organization of outstanding engineers. It is autonomous in its administration and in the selection of its members, sharing with the National Academy of Sciences the responsibility for advising the federal government. The National Academy of Engineering also sponsors engineering programs aimed at meeting national needs, encourages education and research, and recognizes the superior achievements of engineers. Dr. Charles M. Vest is president of the National Academy of Engineering.

The **Institute of Medicine** was established in 1970 by the National Academy of Sciences to secure the services of eminent members of appropriate professions in the examination of policy matters pertaining to the health of the public. The Institute acts under the responsibility given to the National Academy of Sciences by its congressional charter to be an adviser to the federal government and, upon its own initiative, to identify issues of medical care, research, and education. Dr. Harvey V. Fineberg is president of the Institute of Medicine.

The **National Research Council** was organized by the National Academy of Sciences in 1916 to associate the broad community of science and technology with the Academy's purposes of furthering knowledge and advising the federal government. Functioning in accordance with general policies determined by the Academy, the Council has become the principal operating agency of both the National Academy of Sciences and the National Academy of Engineering in providing services to the government, the public, and the scientific and engineering communities. The Council is administered jointly by both Academies and the Institute of Medicine. Dr. Ralph J. Cicerone and Dr. Charles M. Vest are chair and vice chair, respectively, of the National Research Council.

www.nationalacademies.org

OTHER RECENT REPORTS OF THE AERONAUTICS AND SPACE ENGINEERING BOARD

An Interim Report on NASA's Draft Space Technology Roadmaps (Aeronautics and Space Engineering Board [ASEB], 2011)
Limiting Future Collision Risk to Spacecraft: An Assessment of NASA's Meteoroid and Orbital Debris Programs (ASEB, 2011)
Recapturing a Future for Space Exploration: Life and Physical Sciences Research for a New Era [prepublication version] (Space Studies Board [SSB] with ASEB, 2011)
Summary of the Workshop to Identify Gaps and Possible Directions for NASA's Meteoroid and Orbital Debris Programs (ASEB, 2011)

Advancing Aeronautical Safety: A Review of NASA's Aviation Safety-Related Research Programs (ASEB, 2010)
Capabilities for the Future: An Assessment of NASA Laboratories for Basic Research (Laboratory Assessments Board with ASEB, 2010)
Defending Planet Earth: Near-Earth-Object Surveys and Hazard Mitigation Strategies: Final Report (SSB with ASEB, 2010)
Final Report of the Committee to Review Proposals to the 2010 Ohio Third Frontier (OTF) Wright Projects Program (WPP) (ASEB, 2010)

America's Future in Space: Aligning the Civil Space Program with National Needs (SSB with ASEB, 2009)
Approaches to Future Space Cooperation and Competition in a Globalizing World: Summary of a Workshop (SSB with ASEB, 2009)
An Assessment of NASA's National Aviation Operations Monitoring Service (ASEB, 2009)
Final Report of the Committee for the Review of Proposals to the 2009 Engineering and Physical Science Research and Commercialization Program of the Ohio Third Frontier Program (ASEB, 2009)
Fostering Visions for the Future: A Review of the NASA Institute for Advanced Concepts (ASEB, 2009)
Near-Earth Object Surveys and Hazard Mitigation Strategies: Interim Report (SSB with ASEB, 2009)
Radioisotope Power Systems: An Imperative for Maintaining U.S. Leadership in Space Exploration (SSB with ASEB, 2009)

Assessing the Research and Development Plan for the Next Generation Air Transportation System: Summary of a Workshop (ASEB, 2008)
A Constrained Space Exploration Technology Program: A Review of NASA's Exploration Technology Development Program (ASEB, 2008)
Launching Science: Science Opportunities Provided by NASA's Constellation System (SSB with ASEB, 2008)
Managing Space Radiation Risk in the New Era of Space Exploration (ASEB, 2008)
NASA Aeronautics Research: An Assessment (ASEB, 2008)
Review of NASA's Exploration Technology Development Program: An Interim Report (ASEB, 2008)
Science Opportunities Enabled by NASA's Constellation System: Interim Report (SSB with ASEB, 2008)
United States Civil Space Policy: Summary of a Workshop (SSB with ASEB, 2008)
Wake Turbulence: An Obstacle to Increased Air Traffic Capacity (ASEB, 2008)

Limited copies of ASEB reports are available free of charge from

Aeronautics and Space Engineering Board
National Research Council
The Keck Center of the National Academies
500 Fifth Street, NW, Washington, DC 20001
(202) 334-2858/aseb@nas.edu
www.nationalacademies.org/aseb.html

COMMITTEE ON HUMAN SPACEFLIGHT CREW OPERATIONS

FREDERICK D. GREGORY, Lohfeld Consulting Group, Inc., *Co-Chair*
JOSEPH H. ROTHENBERG, SSC, *Co-Chair*
MICHAEL J. CASSUTT, University of Southern California
RICHARD O. COVEY, United Space Alliance, LLC (retired)
DUANE W. DEAL, Stinger Ghaffarian Technologies, Inc.
BONNIE J. DUNBAR, Dunbar International, LLC
WILLIAM W. HOOVER, Independent Consultant
THOMAS D. JONES, Florida Institute for Human and Machine Cognition
FRANKLIN D. MARTIN, Martin Consulting, Inc.
HENRY McDONALD, University of Tennessee at Chattanooga
AMY R. PRITCHETT, Georgia Institute of Technology
RICHARD N. RICHARDS, Boeing Corporation (retired)
JAMES D. VON SUSKIL, NRG, Texas

Staff

DWAYNE A. DAY, Senior Program Officer, *Study Director*
CATHERINE A. GRUBER, Editor
LEWIS GROSWALD, Research Associate
AMANDA R. THIBAULT, Research Associate
DIONNA WILLIAMS, Program Associate

MICHAEL H. MOLONEY, Director, Aeronautics and Space Engineering Board

AERONAUTICS AND SPACE ENGINEERING BOARD

Preface

The United States has been launching astronauts into space for five decades. During that period, the size of the Astronaut Corps has grown and shrunk periodically in accordance with various program demands. The training facilities required to support the Astronaut Corps have also changed over time to meet the new demands placed on it. NASA has retired the space shuttle but is still operating the International Space Station and adopting a new approach to transporting astronauts to low Earth orbit and eventually beyond. These changes are affecting the role and size of the activities managed by the Flight Crew Operations Directorate of NASA's Johnson Space Center and of the ground training facilities and the fleet of training aircraft used by the Astronaut Corps.

In May 2010, the National Research Council (NRC) was asked by NASA to address several questions related to the Astronaut Corps:

1. How should the role and size of the activities managed by the Johnson Space Center Flight Crew Operations Directorate change after space shuttle retirement and completion of the assembly of the International Space Station (ISS)?
2. What are the requirements for crew-related ground-based facilities after the Space Shuttle program ends?
3. Is the fleet of aircraft used for training the Astronaut Corps a cost-effective means of preparing astronauts to meet the requirements of NASA's human spaceflight program? Are there more cost-effective means of meeting these training requirements?

The NRC was not asked to consider whether the United States should continue human spaceflight, nor whether there are better alternatives for achieving the nation's goals than launching humans into space. Rather, the NRC's charge was predicated on the assumption that U.S. human spaceflight would continue.

The NRC was asked in its task to establish requirements for the role and size of the activities managed by the Flight Crew Operations Directorate and the crew-related ground-based test facilities but was not asked to conduct cost analysis of different requirements. That analysis can be performed by NASA as it chooses whether and how to implement any of the recommendations in this report.

In response to the request, the NRC appointed the Committee on Human Spaceflight Crew Operations. The committee held its first meeting at Johnson Space Center in January 2011 and then met in Washington, D.C., in March and finally in Woods Hole, Massachusetts, in May. The committee received input from a diverse group that included NASA's Astronaut Office, NASA Headquarters officials, and representatives of the Federal Aviation

Administration, the commercial airline industry, the Navy Nuclear Propulsion Office, and companies seeking to provide commercial crew access to the ISS. NASA's Astronaut Office responded to the committee's many requests for information by providing extensive details on the office's activities, on the Flight Crew Operations Directorate's training resources, and on the role of high-performance aircraft training in the overall astronaut training portfolio. The committee acknowledges the extensive cooperation provided by the chief astronaut and the staff of the Astronaut Office.

Acknowledgment of Reviewers

This report has been reviewed in draft form by individuals chosen for their diverse perspectives and technical expertise, in accordance with procedures approved by the Report Review Committee of the National Research Council (NRC). The purpose of this independent review is to provide candid and critical comments that will assist the institution in making its published report as sound as possible and to ensure that the report meets institutional standards for objectivity, evidence, and responsiveness to the study charge. The review comments and draft manuscript remain confidential to protect the integrity of the deliberative process. We wish to thank the following individuals for their review of this report:

Michael Bloomfield, Oceaneering Space Systems,
Kenneth D. Bowersox, Space Exploration Technologies,
John E. Boyington, Jr., DRS Technologies,
Daniel C. Brandenstein, United Space Alliance, LLC,
Eileen Collins, U.S. Air Force (retired),
N. Wayne Hale, Jr., Special Aerospace Services,
Scott Pace, Space Policy Institute and George Washington University,
Richard H. Truly, National Renewable Energy Laboratory (retired),
Jim Voss, Sierra Nevada Corporation, and
Steven Weinberg, University of Texas, Austin.

Although the reviewers listed above have provided many constructive comments and suggestions, they were not asked to endorse the conclusions or recommendations, nor did they see the final draft of the report before its release. The review of this report was overseen by Alton D. Slay, Slay Enterprises Incorporated. Appointed by the NRC, he was responsible for making certain that an independent examination of this report was carried out in accordance with institutional procedures and that all review comments were carefully considered. Responsibility for the final content of this report rests entirely with the authoring committee and the institution.

Contents

APPENDIXES

Summary

In May 2010, the National Research Council (NRC) was asked by NASA to address several questions related to the Astronaut Corps. The NRC's Committee on Human Spaceflight Crew Operations was tasked to answer several questions:

1. How should the role and size of the activities managed by the Johnson Space Center Flight Crew Operations Directorate change after space shuttle retirement and completion of the assembly of the International Space Station (ISS)?
2. What are the requirements for crew-related ground-based facilities after the Space Shuttle program ends?
3. Is the fleet of aircraft used for training the Astronaut Corps a cost-effective means of preparing astronauts to meet the requirements of NASA's human spaceflight program? Are there more cost-effective means of meeting these training requirements?

Although the future of NASA's human spaceflight program has garnered considerable discussion in recent years and there is considerable uncertainty about what the program will involve in the coming years, the committee was not tasked to address whether human spaceflight should continue or what form it should take. The committee's task restricted it to studying activities managed by the Flight Crew Operations Directorate or those closely related to its activities, such as crew-related ground-based facilities and the training aircraft.

COMMITTEE APPROACH

To conduct this study, the committee visited NASA's Johnson Space Center (JSC) and sought information from the center's Flight Crew Operations Directorate (FCOD), which is responsible for selecting and training astronauts for future missions, and from the Mission Operations Directorate (MOD), which is responsible for facilities and mission-specific training for flight and ground crews. The committee did not address ground-based human spaceflight facilities or activities that did not directly support ground-based training, such as astronaut exercises and the flight surgeon's office.

The committee was briefed by a former astronaut who held senior management positions in NASA and by NASA Headquarters officials who had responsibility for human spaceflight, including the associate administrator for safety and mission assurance, who explained the role of crew training in NASA's overall approach to safety and mission assurance. In addition, the committee sought input from potential providers of future commercially procured human spacecraft, such as SpaceX and Sierra Nevada Corp., and from the Federal Aviation Administration (FAA), which is responsible for licensing commercial launch vehicles in the United States and is expected to play a role in licensing future commercial human spacecraft. And it received briefings on training practices of the Naval Reactors Program and the commercial airline industry.

The committee received substantial cooperation and assistance from the Flight Crew Operations Directorate, including the chief of the Astronaut Office and members of her staff. The committee assessed the information provided to it by FCOD in presentations and the planning tools used by FCOD and the Astronaut Office. That cooperation made it possible for the committee to explore all aspects of its task. Taking into consideration questions posed by the committee during its meetings and questions submitted directly to the Astronaut Office, in March 2011 the Astronaut Office produced a white paper containing substantial amounts of information directly relevant to the committee's task.[1] Throughout this report, the committee uses graphics provided by NASA in that white paper, which is referred to in the report as "NASA Astronaut Office 2011 White Paper."

SUMMARY FINDINGS AND RECOMMENDATIONS

Question 1—Role and Size of the Flight Crew Operations Activities

NASA's Astronaut Office, which is part of FCOD at JSC, is responsible for managing NASA's Astronaut Corps, which the committee defines as the set of astronauts qualified to fly into space, excluding astronauts who have transitioned to management positions in the agency and are no longer eligible to fly space missions. The Astronaut Corps has been reduced substantially since it reached a peak of nearly 150 in 2000. In May 2011, the Astronaut Corps consisted of 61 persons, and NASA has projected a minimum required Astronaut Corps of 55-60 astronauts through 2016. In 2009, the agency selected a new class of nine astronaut candidates for addition to the Astronaut Corps in 2011. It is expected that the new class will compensate for any attrition and help to ensure long-term sustainment of a skilled U.S. Astronaut Corps.

Although NASA's human spaceflight program and its post-shuttle crew requirements have not been well defined except in terms of the ISS, the committee concluded that the sizing of the Astronaut Corps to meet ISS crew requirements has been well modeled by using as input ISS crew selection, training and flight recovery times, and a post-shuttle force reduction plan. NASA uses a model to predict the minimum staffing requirements and then applies an arbitrary management margin to the result (Figure S.1).

According to a presentation by the chief of the Astronaut Office to the committee during its first meeting, the model produces a theoretical minimum and does not include several real-world constraints, such as mission-required skills mix, temporary or permanent medical disqualification, inability of astronauts returning from a long-duration mission to fly another long-duration mission at the end of the normal 1.5-year recovery period, and the desired pairing of inexperienced and experienced astronauts on new assignments. In light of those unpredictable constraints, a 25 percent margin, as shown in Figure S.1, is factored into the model used to determine the size of the Astronaut Corps that will meet the minimum manifest requirement.

The committee notes that in addition to substantially reducing the size of the Astronaut Corps recently, NASA has reduced the management margin that it applies to its model. The margin, which was 50 percent, was reduced in 2010 to 25 percent, apparently because of budget pressures. According to the chief of the Astronaut Office, "The corps requirements will always be greater than the manifest analysis, and 25 percent may not be enough margin."[2]

The committee notes that new sources of uncertainty have been identified in the human spaceflight program. For example, a relatively new medical condition has been observed among astronauts returning from long-duration space missions: papilledema, a swelling of the optic disk. The condition has led to several Astronaut Corps mem-

[1] NASA Astronaut Office, "Ensuring the Readiness of the Astronaut Corps: A White Paper," NASA Johnson Space Center, Houston, Tex., March 25, 2011.

[2] P.A. Whitson, Astronaut Office, "Presentation to the NRC Committee on Human Spaceflight Crew Operations," presentation to the Committee on Human Spaceflight Crew Operations, January 6, National Research Council, Washington, D.C., 2011, p. 36.

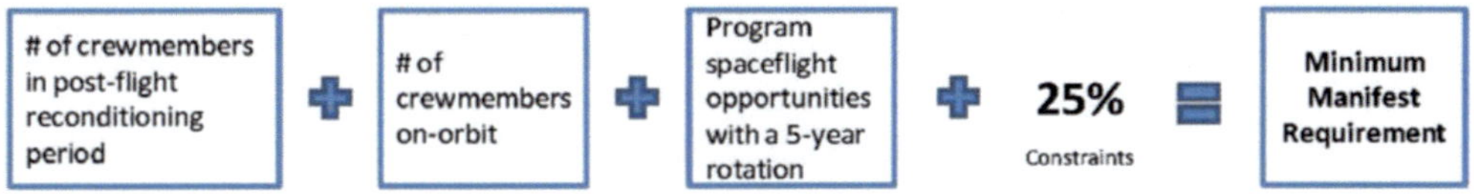

FIGURE S.1 NASA's formula for determining minimum manifest requirements for staffing. SOURCE: NASA Astronaut Office, "Ensuring the Readiness of the Astronaut Corps: A White Paper," NASA Johnson Space Center, Houston, Tex., March 25, 2011.

bers' being medically disqualified from flying again until the condition improves. In the past, attrition rates for the Astronaut Corps were based on rates for both space shuttle and ISS missions. As NASA transitions to only long-duration ISS missions, it is difficult to predict attrition rates. After their first long-duration ISS mission, members of the Astronaut Corps might choose to leave the corps rather than fly another long-duration mission with its attendant stresses on family and home life.

The development of future spacecraft also involves programmatic uncertainties. NASA has traditionally assigned astronauts roles in the development of new vehicles to benefit from their insight regarding design and for reasons of safety and mission assurance. According to NASA, those roles need to be filled by members of the Astronaut Corps who will ultimately fly in the vehicles, not by management astronauts (former astronauts who are no longer eligible for flight assignment and who do not use NASA aircraft or other training facilities except as instructors, evaluators, mentors, or providers of expertise).[3] However, as presented to the committee, NASA's calculations for sizing the Astronaut Corps focused on preparing for planned missions and did not provide for filling those additional roles.

Viewed as a supply chain, astronaut selection and training are sensitive to critical shortfalls because of the long lead times and long recovery time between missions and because astronauts, trained for specific roles and missions, cannot be easily interchanged.

On the basis of its assessment of known and potential needs, the committee concluded that the currently projected minimum target size for the active Astronaut Corps poses a risk to the U.S. investment in human spaceflight capabilities; in particular, the committee notes that the planned Astronaut Corps, sized only to meet ISS crew requirements, would not have the flexibility to accommodate unexpected increases in attrition or commercial, exploration, and new mission development tasks. Because of various sources of uncertainty and because multiple factors are involved in the training of members of the Astronaut Corps and the operation of spacecraft in orbit, it is not possible to quantify the risk posed by tight margins or size or to provide a confidence level of risk. Nevertheless, the committee concluded that the Astronaut Corps is vital to the safe and successful operation of the ISS and that reducing its size too much can create shortages of key skills.

It currently takes 2 years to train a new astronaut in the full range of ISS skills and the fundamentals of spaceflight, aviation, and NASA programs. A newly hired astronaut is not given a technical assignment until about a year after being hired, and training continues. But that new hire will not generally be credible in representing NASA and the Astronaut Corps in, for example, commercial spaceflight, development of beyond-low-Earth-orbit (LEO) spacecraft, or supporting complex ISS operations or emergencies. (New astronauts are, however, highly skilled in their own fields from the outset and could provide immediate expertise in those fields, such as test-piloting, research, and engineering development.)

NASA would be able to respond to long-term programmatic commitments by hiring new classes oriented to them, but the most valuable personnel to retain are those with spaceflight experience and long experience in

[3] Management astronauts can serve as instructors, evaluators, mentors, or providers of expertise and in most cases do not use "additional" training assets. These training assets are being used in the normal course of the event that the astronaut, or whoever would act in that role, is supporting. (In a very small number of cases, additional resources may be required for the astronaut instructor to keep current, but, again, this would be required of any instructor.)

NASA's operations and development programs. FCOD will not have the luxury of hiring new people to deal with a serious failure aboard the ISS or in the early stages of the development and oversight of commercial spacecraft and spacecraft beyond LEO.

NASA's Astronaut Office, which includes both the Astronaut Corps and such additional personnel as management astronauts no longer qualified to fly on space missions, supports several tasks. During its study, the committee noted that the Astronaut Office did not explicitly identify providing operational knowledge and corporate memory of human spaceflight as among its tasks, although this responsibility is implicit in the work that the Astronaut Office does. As a result, the committee specifically identified provision of operational knowledge and corporate memory of human spaceflight as one of the Astronaut Office's tasks, but it notes that maintaining this capability does not drive the minimum manifest requirement for members of the Astronaut Corps.

NASA plans to make periodic selections of a relatively small number of new astronaut candidates over the next few years. The committee believes that that is appropriate and that it is up to NASA to determine how often to make such selections; the committee's recommendation concerns only the model for calculating requirements.

Findings and Recommendations on the Role and Size of the Activities Managed by FCOD[4]

Finding 2.1a. NASA's current Astronaut Office's role is to support six tasks (in priority order):[5]

1. Provide well-trained spaceflight operators to support the NASA flight manifest.
2. Provide ground support personnel for tasks required specifically to support the NASA flight manifest.
3. Provide support for new program development, ranging from development of relatively small payloads and equipment to development of whole new spacecraft designs.
4. Provide operational knowledge and corporate memory of human spaceflight.
5. Provide for collaboration with other government and private organizations as needed and directed by NASA.
6. Provide support for public and educational outreach to society.

The first task is the one in FCOD's model that drives the size of the Astronaut Corps—the number of astronauts qualified to fly in space. But the demands of tasks 2 through 6 add to the workload. The committee supports these roles as a proper use of an important core capability both now and into the future.

Management (inactive) astronauts serving in civil service positions in the Astronaut Office provide supplemental support for tasks 2 through 6. They do not use training assets except as instructors, evaluators, mentors, or providers of expertise; are ineligible for flight; and do not provide a reserve capacity for flight assignments.

Finding 2.1b. Although NASA's human spaceflight program and its post-shuttle crew requirements have not been well defined beyond operation of the ISS, the sizing of the Astronaut Corps to meet ISS crew requirements has been well modeled by using ISS crew selection, training and flight recovery times, and a plan for post-shuttle force reduction.

Finding 2.1c. Astronaut anthropometric (physical size) limitations for flying in the Soyuz limit flexibility in crew assignments in response to contingencies.

Conclusion 2.1. On the basis of its assessment of known and potential needs, the committee concluded that the currently projected minimum staffing target size for the active Astronaut Corps poses a risk to the U.S. investment in human spaceflight capabilities. The committee concluded that given the array of potential crew assignment constraints and uncertainty in future requirements, the Astronaut Corps appears

[4] The numbering of the findings and recommendations mirrors their numbering in Chapters 2 and 3.

[5] NASA identified tasks 1, 2, 3, 5, and 6; the committee has added task 4.

to be sized below the minimum required. The committee notes that the current plan for the size of the Astronaut Corps does not have the flexibility to accommodate commercial, exploration, and new mission development tasks or unexpected increases in attrition.

Recommendation 2.1.

- The committee recommends that the factor for uncertainty used by the Astronaut Office in its model to determine minimum staffing requirements for the Astronaut Corps be increased above the current 25 percent, which is inadequate to provide sufficient flexibility to meet the current flight manifest requirements reliably.
- In addition to task 1, the Astronaut Office should maintain the staff required to accomplish tasks 2 through 6 as listed in Finding 2.1a.

Finding 2.2. In addition to the need to meet NASA requirements, there is an expectation on the part of commercial crew providers and the Federal Aviation Administration that FCOD expertise and capabilities will be available in the future.

Recommendation 2.2. NASA's Flight Crew Operations Directorate should continue to serve as a national resource for U.S. human spaceflight experience and knowledge. This resource should be

- Maintained to ensure appropriate staffing and training of the Astronaut Corps in support of the International Space Station manifest;
- Applied to the future development of NASA human spaceflight and exploration activities;
- Available to the emerging commercial space industry and the FAA; and
- Applied to support authorized agreements with international partners.

Question 2—Ground-Based Training Facilities

The committee assessed the requirements for ground-based training and simulation facilities, such as the Neutral Buoyancy Laboratory, the ISS part-task trainers, and the shuttle simulators. It found that the NASA plan for the size and tasks of the Astronaut Corps to support the on-orbit and ground requirements for the ISS will use both high-performance aircraft maintained by the JSC FCOD and ISS mission-specific ground facilities maintained by the JSC MOD. The committee found that after space shuttle retirement the mission-specific spaceflight operations requirements for crew will shift from shuttle operations and ISS assembly to Soyuz and ISS operations, ISS maintenance, and emergency response. The committee believes that to accomplish training for those tasks, more specifically to be able to respond to on-orbit problems safely and successfully, NASA will need to maintain the current ISS training facilities. As the requirements for commercial crew support and the future human space exploration program become clearer, there may be a need for additional astronauts and new flight or ground-based training facilities.

Findings and Recommendations on Ground-Based Facilities

Finding 3.1. The NASA plan for post-shuttle retirement of shuttle-specific training facilities is generally appropriate. However, the Shuttle Engineering Simulator Dome may be useful in training for future activities, such as rendezvous and docking operations during commercial transportation of ISS crew.

Recommendation 3.1. NASA should evaluate potential future requirements for the Shuttle Engineering Simulator Dome and, if it will be needed, should preserve this facility.

Finding 3.2a. Now that the shuttle is retired, the specific spaceflight crew operations shift from shuttle operations and ISS assembly to Soyuz and ISS nominal and emergency operations, ISS payload operations, and ISS maintenance. The requirements for training of flight crews for those ISS operations include emergency response training, extravehicular activity operations, and the full suite of nominal operations for U.S. and international partner ISS elements, including Soyuz. Thus, the ISS ground-based training facilities are required for the support of crew training for future operations and maintenance of the ISS.

Finding 3.2b. The requirements for U.S. astronaut training include international partner ISS element operations at international partner facilities and Soyuz operations in Russia. The U.S. international partner agreements also require that the United States provide for enhancing skill proficiency and training for the international partner astronauts.

Recommendation 3.2. NASA should retain the capability and training facilities to conduct International Space Station (ISS) mission-specific training after retirement of the space shuttle to ensure the continued safety and mission success of ISS operations.

Question 3—Aircraft Training

Ground-based simulators for spaceflight missions are used for approximately 90 percent of crew training. Only a small part of the training, designated by NASA as spaceflight readiness training (SFRT), puts the crew into operational environments in which they share some aspects of the fast dynamics, physical stress, and risk found in spaceflight. This kind of training is currently accomplished primarily by using a T-38N Talon two-person jet, a high-performance training aircraft that was originally purchased by the Air Force as a fighter pilot trainer more than 50 years ago. (Flight in the T-38N represents 10 percent of the training time for unassigned astronauts and 5 percent of the training time for those assigned to an upcoming mission. The most significant constraint on assigned astronauts' acquiring flight time is that they are training overseas for a great deal of time.)

It is important to emphasize that SFRT is not just about flying the T-38N as the pilot in command. Instead, it is about developing the skills and ability to work together in an environment that is fast-paced, is physically stressful, and carries potentially severe penalties for failure. SFRT involves both the pilot in command of the aircraft and the person in the backseat, dividing responsibilities. For example, the backseat flyer frequently handles navigation, communications, and crew resource management duties during flight and must coordinate with the pilot, who is actually flying the aircraft. Hands-on control of the aircraft by backseaters is a big part of SFRT. SFRT is useful for many aspects of spaceflight—not only for operation of a spacecraft, such as Soyuz, but for operations onboard the International Space Station.

NASA currently has a fleet of 21 T-38s for astronaut training and intends to reduce this number to 16 in 2013 because of a planned reduction in the size of the Astronaut Corps. Small "environmental" additions to flight training include exposure to hypoxia in the JSC high-altitude chamber, a one-time run in the Russian centrifuge at Star City[6] (for a medical evaluation during exposure to ballistic re-entry), and survival training (for water and land). SFRT is designed to provide evaluation of training and crew performance in an environment that most closely replicates the physical and psychological stresses of the high-speed dynamic environment of spaceflight.

The committee assessed the value of using high-performance training aircraft, the basis of the projected size of the current T-38N fleet and alternative aircraft, and the potential for using high-fidelity simulators for SFRT. The committee also was provided input from the Navy Nuclear Propulsion Program regarding its training tools and experiences and from the commercial aviation industry.

Findings and Recommendations on Spaceflight Readiness Training

Finding 3.3a. The spaceflight readiness training requirement is derived from safety and mission success requirements, not tied to any specific mission. Although the requirement is not expressly documented at the NASA Headquarters program level, it was developed by the Flight Crew Operations Directorate

[6] The most accurate translation of the Russian name is "Starry Town," but "Star City" is the common usage at NASA.

in response to NASA Headquarters-controlled safety and mission success requirements and embedded at the level of the NASA JSC Certificate of Flight Readiness for safe operations of flight, which is then provided to NASA Headquarters. Any changes in spaceflight readiness training need to be made with great care because changes can result in increased risk to safety and mission success.

Finding 3.3b. Spaceflight readiness training using high-performance aircraft has been demonstrated and documented to prepare crews for successful and safe spaceflight, dating back 50 years, from the inception of the Mercury program to the current International Space Station program. SFRT is more than just flying—the full spectrum of experiences gained is not restricted to the operation of high-performance aircraft but extrapolates to crew resource management and performance under stress. SFRT is used effectively internationally to produce qualified members of the Astronaut Corps who are independent of crew position or vehicle design.

Recommendation 3.3. To ensure continued safety and mission success, NASA should maintain a spaceflight readiness training program that includes high-performance aircraft.

Finding 3.4a. FCOD maintains the Astronaut Corps and provides the capability to conduct SFRT.

Finding 3.4b. High-performance aircraft present conditions, including crew disorientation and rapid fluctuation in G-forces, under which the flight crew must carry out complex tasks in a stressful and potentially life-threatening environment. That combination of unique environments, demand for rapid, critical decision making, and historical evidence convinced the committee that SFRT provides experience-based training that cannot be duplicated by current or, to the best of the committee's knowledge, projected alternative techniques or technologies.

Finding 3.4c. Given the current investment in the existing T-38N fleet, this fleet is the most cost-effective means of providing SFRT in the near term. In the long term, new technology that may be a more cost-effective means of providing SFRT might be demonstrated and proved.

Finding 3.4d. The size of the T-38N SFRT fleet is projected to fall to 16 aircraft in 2013.

Recommendation 3.4. NASA should retain the T-38N fleet for spaceflight readiness training and should fund the fleet at a level commensurate with the projected required size of the post-shuttle Astronaut Corps.

Finding and Recommendation on Learning from Other Occupations

Finding 3.5. Substantial research is being undertaken on selection and training of personnel in related high-stress occupations. Some of that work is leading to continually improving methods and technologies for training for team and individual performance in stressful high-risk situations.

Recommendation 3.5. NASA should continue to monitor training methods and technologies in related fields for possible ways to enhance the astronaut selection and training process.

ORGANIZATION OF THIS REPORT

Chapter 1 of this report explains the evolution of the U.S. astronaut program, detailing how the Astronaut Corps has grown and shrunk over time, why it has done so, and how its composition has changed. It also describes the evolution of the ground training facilities. Finally, it explains how and why spaceflight readiness training that requires an aircraft training capability was introduced to the program and how that capability has evolved.

Chapter 2 responds to the committee's first task. It explains the current status of NASA's FCOD, its Astronaut

Office, and the Astronaut Corps. It also explains new situations that the human spaceflight program is facing, such as the introduction of new spacecraft. It provides findings and recommendations concerning the role and size of the FCOD's activities and the size of the Astronaut Corps.

Chapter 3 responds to the committee's second and third tasks regarding ground-based training facilities and the T-38N fleet. It explains the role of spaceflight readiness training, how it is served by the T-38N fleet of training aircraft, and possibilities for future simulator-based training; it also notes that other occupations offer potential lessons for NASA astronaut training. It provides findings and recommendations concerning ground-based training facilities, the T-38N aircraft fleet, and simulators.

1

The Evolution of the U.S. Astronaut Program

In the half century since the flight of Yuri Gagarin, more than 500 people have orbited Earth, of which 24 traveled to the Moon. Approximately 61 percent have been Americans.

With the exception of a handful of self-funded spaceflight participants, Soviet Union guest cosmonauts, and U.S. shuttle-era payload specialists, those astronauts have all traveled a relatively common road through the mission selection process. Each was subjected to rigorous medical and psychological screening, basic training in common academic classroom technical subjects, systems training in the spacecraft being flown, integrated emergency procedures, and crew-related team building exercises. They also transferred their technical knowledge and training into the design of the vehicles they flew and into the orbital operation of complex scientific and engineering research equipment and experiments. They have been considered not just operators but integral participants in the development and testing of new technologies and vehicles. In addition, they were exposed to both psychologically and physically stressful environments during training and in the flight environment. Before venturing to the launch pad, each professional would-be astronaut or cosmonaut endured survival training—the jungles, mountains, or water—and some mastered the art of parachute jumping or logged hundreds of hours of flight in high-performance aircraft, all with the goal of being able to perform in highly stressful and unusual environments.

Since 1961, U.S. astronauts have trained on and flown seven spacecraft systems, walked on the Moon, assembled a space station, retrieved satellites, launched and repaired the Hubble Space Telescope and other satellites, and trained on and executed thousands of scientific and engineering research experiments. Now, with the end of the Space Shuttle program and its unique training requirements, the NASA Johnson Space Center FCOD and the Mission Operations Directorate (MOD) Training Division are reviewing astronaut staffing and training facilities for the future. Future requirements for support of the International Space Station (ISS) are being coordinated with the ISS international partners, but commercial spaceflight and space exploration beyond low Earth orbit remain undefined. This transitional period creates uncertainty and challenges for NASA in determining the best staffing size for the Astronaut Corps.

HISTORY OF THE SIZE OF THE NASA ASTRONAUT CORPS

The number of astronauts qualified to fly in space as part of the Astronaut Corps has varied primarily as a function of the active flight program flight rate and vehicle crew size capability (Table 1.1). In the early 1960s, the Astronaut Corps started with seven astronauts during the Mercury program and grew to a high of nearly 150

TABLE 1.1 U.S. Human Spaceflight Programs

Program Name	Vehicle Development	Flight Operations	Number of Flights	Crew Size
Mercury	1958-1961	1961-1963	6	1
Gemini	1958-1965	1965-1966	10	2
Apollo	1960-1968	1968-1975	12	3
Skylab Space Station[a]	1966-1973	1973-1974	3	3
Space shuttle	1972-1981	1981-2011	135	2-8
Space shuttle-Mir[b]		1995-1998	9	7+
International Space Station[c]	1993-2011	1998-TBD	27	3-6
Constellation Exploration[d]	2004-2011	Canceled	0	
Orion/multipurpose crew vehicle	2005-?	NLT 2016?	0	4

[a] First long-duration missions for the United States. Longest of 84 days set world record and established high standard of biomedical data for long-duration missions.

[b] Docking between the space shuttle and the Russian Space Station, Mir. Flight of U.S. astronauts on Russian Soyuz to Mir and Russian cosmonauts onboard the space shuttle. Seven astronauts were left on the Mir for long-duration missions (greater than 90 days). Program used existing Astronaut Corps (no new selections) but required new vehicle training on Russian Soyuz and Mir Space Station and Russian language training. In addition, shuttle crews were trained in docking operations with Mir. All Russian vehicle training was executed at Gagarin Cosmonaut Training Center in Star City, Russia.

[c] Flight is defined as "increment," which is a multimonth mission and separated from the vehicle used to reach or return from the ISS.

[d] The Exploration Program to the Moon and Mars was canceled in 2010, but the Astronaut Office had been closely involved in requirements and design.

at the peak of space shuttle flights and preparation for the ISS in 2000. Vehicle habitable volume and flight rate also increased. The Mercury capsule flew with a crew of one, and the later space shuttle could accommodate a maximum crew of eight. The ISS hosts an international crew of six. The current size of the active U.S. Astronaut Corps is 61, and an additional nine astronauts are in training (astronaut candidates, referred to as ASCANs). NASA has projected a minimum required Astronaut Corps size of 55 to 60 astronauts through 2016.

The size of the Astronaut Corps has historically been aligned not just with the spacecraft being flown at the time but with future human spaceflight programs in development but not yet flying. Because of the lead time required to train an astronaut through a basic and then a mission-specific training syllabus—often up to 4 years for the ISS—the Astronaut Office and the Mission Operations Training Division were required to develop reliable forecasting algorithms for personnel and facilities. The forecasts also typically tried to accommodate anticipated astronaut attrition due, for example, to retirements, large gaps in flight opportunities, health issues, and, more recently, lifetime radiation limits and Russian anthropometric size requirements.

The U.S. human spaceflight program history, from program development to flight operations, is summarized in Table 1.1. The table does not reflect the current dependence on the Russian Soyuz capsule (three cosmonauts) to reach the International Space Station after shuttle retirement. With retirement of the space shuttle, flights of U.S. astronauts to the ISS will decrease from about 28 a year to fewer than 6. However, the time to train for a mission increases from approximately 1 year for a shuttle to 3-4 years for the ISS. Furthermore, individual astronauts will not be able to fly as frequently on the ISS as on the space shuttle, because of lifetime radiation dose limits currently imposed on them.

CREW REDUNDANCY AND BACKUPS

The Mercury, Gemini, and Apollo programs required that a full backup crew be trained for each flight. The backup astronauts were required when medical problems, or even death, affected the crew manifest. The early schedules were driven by the Cold War space race, so resources were available to support the missions fully.

Backup crew members were used in the early flights of the space shuttle (STS-1 through STS-4), when the crew was limited to two. However, when the flights expanded to four and later to five and then eight, it was decided to cross-train on the same flight for many of the functions because there were not enough astronauts and training resources to fully provide for backup crews. Two pilots with dual shuttle controls provided redundancy in space shuttle landing, which turned out not to have been needed. Mission specialists were generally cross-trained for on-orbit operations, with two notable exceptions: the flight engineer position (MS2) and for extravehicular activity (EVA), which required suits of a particular size. Each space shuttle flight required at least two astronauts qualified for EVA. EVA is also a requirement on the ISS. During this period, if an astronaut had to be replaced, it was assumed that an astronaut who had flown would be drawn from those awaiting their next flight assignment (Figure 1.1).

In addition, the composition of the Astronaut Corps has evolved substantially as well, and the current corps is a technically diverse group of people who have widely varied backgrounds and experience. The astronaut program began with only military test pilots, progressed to include scientist-astronauts during Apollo and Skylab, and now is composed of test pilots and mission specialists (engineers, scientists, and physicians) and educators and international partner astronauts sent by their home agencies to train alongside their U.S. colleagues.

The major changes in the professional Astronaut Corps have corresponded with the introduction of new spacecraft designs, with the addition of research to the mission objectives, and with the adoption of new policy goals for the nation's human spaceflight programs. The Space Shuttle program, by design, was opened to women and a more technically diverse set of engineers and scientists (mission specialist astronauts and payload specialists).

FIGURE 1.1 The first U.S. spacewalk was performed by Ed White during the Gemini 4 mission on June 3, 1965. The addition of new tasks, such as spacewalks, increased astronaut training requirements. SOURCE: Courtesy of NASA.

The space shuttle also opened up the anthropometric limitations for all potential astronaut applicants and allowed larger men (95th percentile) and smaller women (5th percentile) to be qualified for selection.

Beginning with the Space Shuttle program, additional crew members have trained with the professional astronauts. They included payload specialists—career scientists and engineers who typically flew only once and then returned to their home laboratories or companies. They were often principal investigators and trained typically for more than 18 months on the mission research manifested with the Spacelab but for only 6 months with the crew and space shuttle systems. Some were corporate researchers. Eventually, this category would be expanded to include spaceflight participants, who were general observers of the spaceflight experience. Approximately 60 payload specialists have flown with the Space Shuttle program.

Table 1.2 shows 20 selected and announced classes of astronauts from 1959 to 2009, a span of 50 years: 148 pilots, 17 scientist-astronauts (Apollo era), 163 mission specialists (space shuttle to present), 35 international partner astronauts (Japan, Europe, and Canada), and three educators. All were trained in the training facilities at the NASA Johnson Space Center.

Figure 1.2 illustrates the historical Astronaut Corps size and trends from 1959 to the present. Several trends

TABLE 1.2 Astronaut Class Composition from 1959 to 2009

Year	Class #	Pilots	Apollo and Skylab Scientists	Mission Specialists	International Partners	Educators	Total
1959	1	7	0	0	0	0	7
1962	2	9	0	0	0	0	9
1963	3	14					14[a]
1965	4		6				6[b]
1966	5	19					19[c]
1967	6		11				11[d]
1969	7	7					7
1978	8	13		22			35
1980	9	8		11	2		21[e]
1984	10	7		10			17
1985	11	6		8			14[f]
1987	12	7		8			15
1990	13	7		16			23
1992	14	4		15	5		24[g]
1995	15	10		9	4		23
1996	16	10		25	9		44
1998	17	8		17	7		32
2000	18	7		10			17
2004	19	2		6	3	3	14[h]
2009	20	3		6	5		14
Total		148	17	163	35	3	366

[a] Four died in training accidents before they could fly.

[b] Four had prior military experience. Two left NASA without having flown in space. All had delayed flight assignments because of the requirement that they spend a year at U.S. Air Force (USAF) Undergraduate Pilot Training (UPT) to be jet pilot qualified.

[c] All with military experience. One died in an accident before flight, one left because of illness before flight.

[d] All had delayed flight assignments because of the requirement that they spend a year at USAF UPT to be jet pilot qualified. Seven remained after Apollo and formed the core of the space shuttle mission specialists before the 1978 astronaut selection. Four did not complete training for flight.

[e] First two European Space Agency (ESA) astronauts to be assigned to train as mission specialists. Trained for first year and then returned to ESA for payload training. One was professional commercial and military pilot.

[f] First teacher in place, Christa McAuliffe, assigned but trained in "payload specialist" curricula, which generally started 6 months before launch. Number not counted in total.

[g] International partner astronauts announced with NASA classes. Participated as mission specialists, including T-38N training, before flight.

[h] Educators trained as full mission specialists, including T-38N spaceflight readiness training. Does not include the Russian cosmonauts, who announce their own classes, even though they are part of future joint ISS and Soyuz crews.

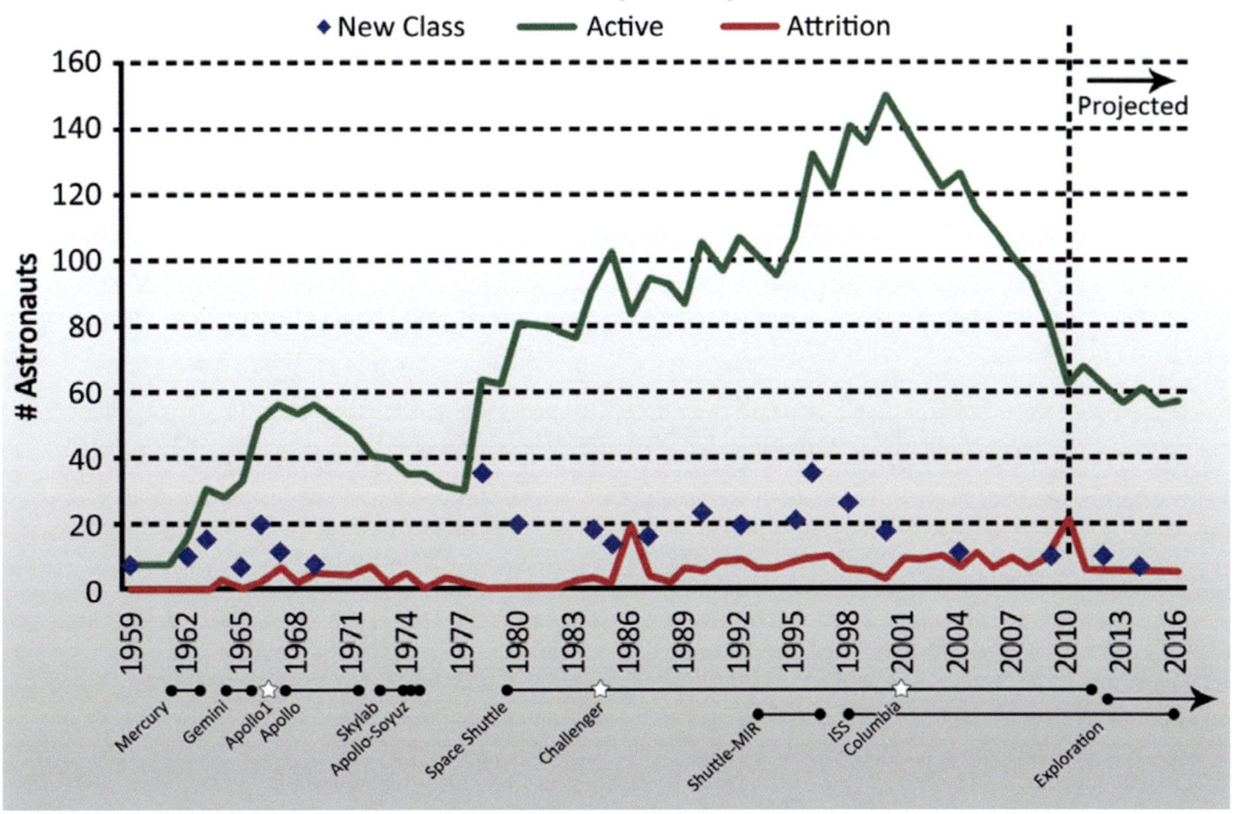

FIGURE 1.2 Historic Astronaut Corps population, December 15, 2010. SOURCE: NASA Astronaut Office, "Ensuring the Readiness of the Astronaut Corps: A White Paper," NASA Johnson Space Center, Houston, Tex., March 25, 2011.

are notable: (1) the decrease in office size during the flight gap between Apollo and the space shuttle, (2) the increase in size with the advent of the shuttle and the ISS, (3) the decrease after the *Challenger* accident with its flight gap, and (4) the steady decline starting in 2000 as a result of policy decisions and an understanding that the ISS would require fewer astronauts over longer periods.

HISTORY OF ASTRONAUT CORPS SELECTION CRITERIA AND AVIATION EXPERIENCE

When they began their human spaceflight programs, the United States, the Soviet Union, and China all initially selected crews from their aviation populations, primarily high-performance jet aircraft. They initially did that for skill screening reasons, and all three countries then sought to maintain that proficiency for a variety of skill and safety reasons. The United States and Russia, however, have substantially expanded the population from which they select astronauts or cosmonauts to include those who have little or no aviation experience. Russian cosmonauts now include design engineers from Energia, the primary contractor for its spacecraft. The U.S. program includes a wide variety of scientists, engineers, and physicians.

Before Apollo, President Dwight D. Eisenhower and his advisers considered many occupations for the first astronauts, including mountain climbers, deep sea divers, and other physically risky occupations. Test pilots were chosen because their operationally fast-paced flight environment (requiring rapid and critical decision making skills) and man-machine interface skills (working in an enclosed multitasking environment) were most applicable to the rigors anticipated in spaceflight. Test pilots were also researchers who provided more than piloting skills: they could analytically assess the performance of new vehicles. It was also determined that psychologically and from a skill point of view they had already been evaluated by the military, and this would shorten the overall training time. To a large extent, those selected to become astronauts had already been screened to determine whether they could perform in an intense and hazardous flight environment. The final determination to select military test pilots was made by President Eisenhower, notably in large part because of potential classified aspects of the program.[1]

Mercury

The first members of the NASA Astronaut Corps (Group I, Mercury 7) were all active-duty military test pilots or graduates of the Air Force or Navy test pilot school who had at least 1,500 hours of flying time. The Mercury astronaut training program was divided into six major topic sections:

> The primary requirement, of course, is to train the astronaut to operate the vehicle. In addition, it is desirable that he have good background knowledge of such scientific areas related to space flight as propulsion, trajectories, astronomy, and astrophysics. He must be exposed to and familiarized with the conditions of space flight such as acceleration, weightlessness, heat, vibration, noise, and disorientation. He must prepare himself physically for those stresses which he will encounter in space flight. . . . An aspect of the training which might be overlooked is the maintenance of the flying skill which was an important factor in his original selection for the Mercury program.[2]

The Mercury astronauts, in addition to classroom training and travel to all the spacecraft development sites, participated in centrifuge training, flew on the parabolic aircraft, and flew high-performance aircraft. In the vernacular of the time, "high performance" was defined as jet powered and capable of supersonic speeds, aerobatics, and high-G loads or variable acceleration. Mercury astronauts also helped to test their own pressure suits in various thermal environments and under reduced pressure. (See Figures 1.3 and 1.4.)

In 1960, Lieutenant Robert Voas, a U.S. Navy Medical Service Corps officer assigned to NASA's Space Task Group during the Mercury program, explained the aviation requirements for astronaut training in the early years of NASA's human spaceflight program:

> One of the continuing problems in training for space flight is the limited opportunity for actual flight practice and proficiency training. The total flight time in the Mercury capsule will be no more than 4 to 5 hours over a period of 3 years for each astronaut. The question arises as to whether all the skills required in operating the Mercury vehicle can be maintained purely through ground simulation. One problem with ground simulation relates to its primary benefit. Flying a ground simulator never results in injury to the occupant or damage to the equipment. The penalty for failure is merely the requirement to repeat the exercise. In actual flight operations, failures are penalized far more severely. A major portion of the astronaut's tasks involves high level decision making. It seems questionable whether skill in making such decisions can be maintained under radically altered motivational conditions. Under the assumption that vigilant decision making is best maintained by experience in flight operations, the Mercury astronauts have been provided with the opportunity to fly high-performance aircraft. The program in this area is a result of their own interest and initiative and is made possible by the loan and maintenance of two F-102 aircraft by the Air Force. . . .
>
> Finally, it seems important to reiterate the requirements for reproducing adequate motivational conditions in the training program. The basic task of the astronaut is to make critical decisions under adverse conditions. The results of the decisions he makes involve not just minor discomforts or annoyances, but major loss of equipment and even survival. Performance of this task requires a vigilance and decision making capability difficult to achieve under the

[1] "The inherent riskiness of space flight and the potential national security implications of the program, pointed toward the use of military personnel. It also narrowed and refined the candidate pool, giving NASA a reasonable starting point for selection. It also made good sense in that NASA envisioned the Astronaut Corps first as pilots operating experimental flying machines, and only later as working scientists." John Logsdon and Roger Launius, eds., *Exploring the Unknown, Volume VII, Human Spaceflight: Projects Mercury, Gemini, and Apollo*, NASA, 2008, p. 13.

[2] J. Logsdon and R. Launius, eds., *Exploring the Unknown, Volume VII, Human Spaceflight: Projects Mercury, Gemini, and Apollo*, NASA, Washington, D.C., 2008.

FIGURE 1.3 Mercury capsule 2 pictured at what was formerly Lewis Hangar (now Glenn Research Center) in Cleveland, Ohio. Mercury required only a single astronaut. Later spacecraft could carry more astronauts and increased the need for astronauts. SOURCE: Great Images in NASA; GPN-2000-000382; courtesy of NASA.

FIGURE 1.4 The Mercury 7 astronauts pose beside an F-106B Delta Dart. SOURCE: Courtesy of NASA; GPN-2000-001286, available at http://grin.hq.nasa.gov/.

> artificial conditions of ground simulation. It appears probable that training in ground devices should be augmented with flight operations to provide realistic operational conditions.[3]

When the Astronaut Corps reached 24 active pilots, the Manned Spacecraft Center recognized the need for a regular squadron of planes to keep the crew flying skills sharp.[4] Astronauts were flying in a limited fleet of T-33s and F-102s, but NASA finally narrowed down the selection of permanent aircraft to the F-4 Phantom and the T-38 Talon. They decided on the T-38 because of cost and the fact that it was a well-supported U.S. Air Force (USAF) trainer. The Air Force loaned NASA 5 T-38s in early 1964, but NASA quickly purchased a fleet of 25 T-38s. The fleet of T-38 aircraft was maintained by NASA at the former Ellington Air Force Base near Johnson Space Center. The aircraft were eventually used by both the professional pilot astronauts and the scientist-astronauts to introduce and maintain skills deemed necessary for successful and safe spaceflight; this activity became known as spaceflight readiness training (SFRT).

[3] R.B. Voas, NASA Space Task Group, "Project Mercury Astronaut Training Program," May 30, 1960, Document I-31, pp. 161-172, in *Exploring the Unknown, Volume VII, Human Spaceflight: Projects Mercury, Gemini, and Apollo* (J. Logsdon and R. Launius, eds.), NASA, Washington, D.C., 2008.

[4] D. Slayton and M. Cassutt, *Deke!*, Forge Books, Tom Doherty, Inc., New York, N.Y., 1995, p. 142.

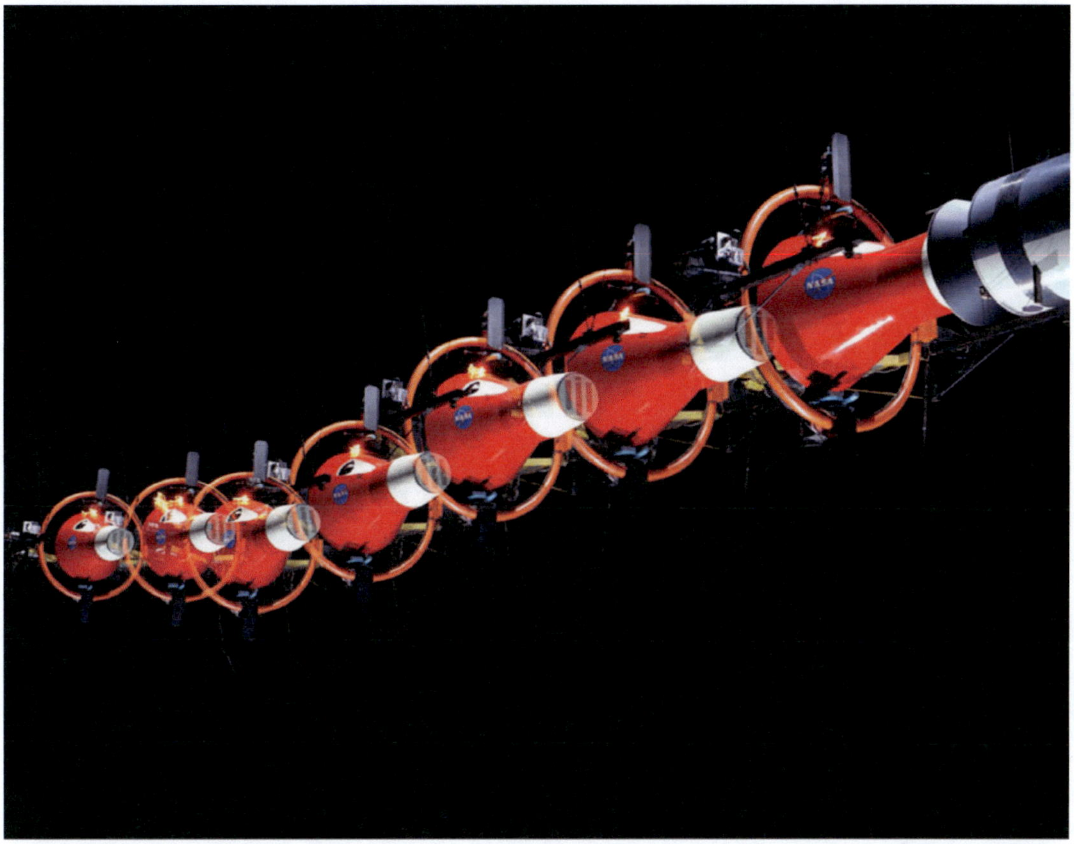

FIGURE 1.5 The Gemini program not only increased the number of astronauts that were required for each spacecraft but added new operations, such as rendezvous, that increased training requirements. Here is a multiple exposure of the Gemini Rendezvous Docking Simulator. SOURCE: Courtesy of NASA/William Salyer; GPN-2000-001278, available at http://grin.hq.nasa.gov/.

Gemini and Apollo

With the approval of the Apollo program in May 1961 and the projected need for astronauts who would fly the missions (and play a lead role in hardware development and in the development of new procedures, such as rendezvous and docking), the agency returned to the test pilot pool for its next selection; this time, however, civilians, including NASA employees, were allowed to apply. Nine new astronauts were selected in September 1962 (Group II), bringing the total Astronaut Office size to 16. (See Figures 1.5, 1.6, 1.7, and 1.8.)

NASA preferred to select experienced test pilots as astronauts, not just for their proven ability to operate in high-stress environments but for their engineering expertise in high-tech development programs. Group II included two civilians who had military test flying experience, four men from the Air Force, and three from the Navy. Three had master's degrees in engineering, four had bachelor's degrees in engineering, and two had bachelor's degrees in science from the Naval Academy.

Believing that the program needed an Astronaut Corps of 24 to crew Gemini and Apollo, NASA management conducted a new recruitment in 1963 (Group III) that gave preference to test pilots but allowed for the selection of candidates with only 1,000 hours of high-performance jet flying time. Fourteen new astronauts were announced in October 1963, bringing the total Astronaut Office head count to 30. Group III included eight holders of advanced degrees and two civilians.

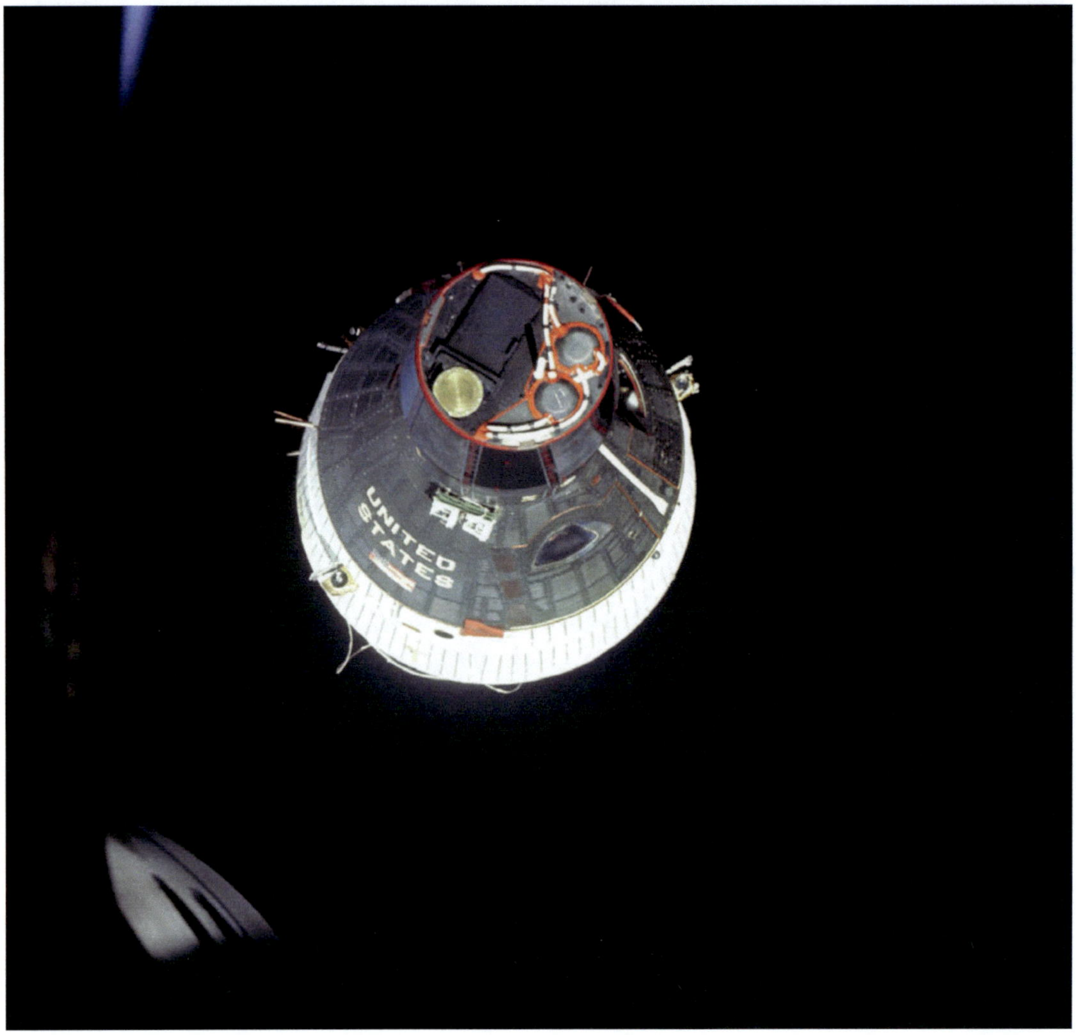

FIGURE 1.6 The Gemini 7 spacecraft as observed from the Gemini 6 spacecraft during rendezvous maneuvers. SOURCE: Courtesy of NASA; GPN-2000-001049, available at http://grin.hq.nasa.gov/.

Scientist-Astronauts and the Apollo Applications Program

With encouragement from the National Academy of Sciences and other outside agencies, beginning in 1964, NASA began to consider the selection of scientists as astronauts for Apollo lunar exploration and Earth orbital missions. Candidates for the scientist-astronaut position could be but were not required to be pilots. The candidates were required to hold a doctoral degree in medicine, engineering, or one of the natural sciences. The National Academy of Sciences screened and evaluated the applications for scientific criteria, and NASA made the final selection.

Of the six candidates selected in June 1965 (Group IV), two were experienced pilots. The other four, as planned, were sent to USAF undergraduate pilot training for a year to qualify in the T-38, which NASA had adopted as its primary aircraft. One candidate withdrew from the program during flight school.

In 1965 and 1966, NASA's long-term schedule called for as many as 30 Apollo Extension Program (AEP) or Apollo Applications Program (AAP) missions from 1969 to the middle 1970s. The agency therefore conducted a new round of selections in two phases: First, a new group of pilot-astronauts were selected according to the same

FIGURE 1.7 Technicians make final inspections of the Gemini 3 spacecraft in the spacecraft preparation room at Kennedy Space Center. SOURCE: Courtesy of NASA; GPN-2006-000016, available at http://grin.hq.nasa.gov/.

FIGURE 1.8 NASA's Lunar Landing Research Vehicle during an early test. New programs and spacecraft required the development of new ground and flying simulators. SOURCE: NASA Dryden Flight Research Center; ECN-448, available at http://nix.ksc.nasa.gov/.

criteria as the October 1963 class. Nineteen (Group V) were announced in April 1966 and included 11 holders of advanced degrees.

A second group of 11 scientist-astronauts (Group VI), selected according to the same standards as the June 1965 group but with NASA taking a more active role in the early screening, were announced in August 1967. In spring 1968, they commenced flight school. It was the first class to contain no military astronauts. Two dropped out during this phase, and the remaining nine reported back to Houston in summer 1969.

During the later Apollo program, after scientist-astronauts were selected (at the insistence of Congress), all non-pilot-astronauts were sent to Air Force flight school for 1 year. When skills related to scientific research, flight engineering, or flight medicine were required and the skilled people did not have aviation experience, NASA management believed that it was necessary for them to acquire the skills and learn how to function in a realistic operational environment. (See Figures 1.9, 1.10, 1.11, 1.12, and 1.13.)

Flight school was also, in itself, part of the selection process. Specifically, not all scientist-astronauts selected for the Astronaut Office and sent to flight school decided to continue in the Corps after that experience.

Apollo 1 Fire in 1967 and Implications for the Astronaut Corps

On January 27, 1967, the Apollo 1 capsule caught fire on the pad at the Kennedy Space Center. Astronauts White, Grissom, and Chaffee perished. Immediately after, NASA convened the Apollo 204 Accident Review

FIGURE 1.9 Apollo 13 astronauts practice a moonwalk at Kennedy Space Center. SOURCE: Courtesy of NASA; GPN-2002-000053, available at http://grin.hq.nasa.gov/.

FIGURE 1.10 Astronaut Alan L. Bean near a tool carrier during an EVA on the Ocean of Storms. SOURCE: Courtesy of NASA/ Charles Conrad, Jr.; GPN-2000-001428, available at http://grin.hq.nasa.gov/.

Board. One result of the accident was that the Astronaut Office became even more integrated into the design and development of the Apollo vehicles (Figure 1.14).

The Astronaut Office size, on July 20, 1969, when the first lunar landing occurred, was 57, more than sufficient for the projected and budgeted Apollo and post-Apollo missions. In late 1969, despite possibly possessing more astronauts than available flight opportunities, NASA agreed to an Air Force request to accept astronauts from the recently canceled Manned Orbiting Laboratory (MOL) program. Seven of the 14 men—all military test pilots—joined the Astronaut Corps, bringing its total size in late 1969 to 64 (Group VII).

Over the next 6 years, as Apollo 13-17 and Skylab 2-4 flew, the Astronaut Corps began to shrink in size: by 1976, a year after Apollo-Soyuz and at least 6 years before the shuttle flights, there were fewer than 30 active astronauts. Most of the attrition was due to astronauts' completing their missions and moving on to other careers, but some astronauts never flew in space and left for various reasons.

FIGURE 1.11 Astronaut Eugene A. Cernan of Apollo 17 salutes the flag on the Moon during an EVA. SOURCE: Courtesy of NASA/Harrison Schmitt; GPN-2000-001273, available at http://grin.hq.nasa.gov/.

The Shuttle Era: 1978-1986

By 1972, when the space shuttle contract was awarded to Rockwell International, NASA began planning for a new type of astronaut to fly on and operate the space shuttle. A NASA selection panel developed criteria for two types of NASA astronauts: pilots and mission specialists. Pilots would serve as commanders and pilots for shuttle missions. They could be military or civilians. Test pilot qualification was not mandatory but was preferred. Mission specialists would operate the Remote Manipulator System (RMS, the shuttle arm) and conduct extravehicular activities, scientific and engineering research, and other non-flying tasks. Mission specialists could be engineers, scientists, or physicians; they could be civilian or non-rated (nonpilot) military personnel (military mission specialists were pilots who lacked the experience for the shuttle pilot position, flight engineers, Navy divers, and so on).

During this period, NASA made an effort to recruit women and minority-group members. Both the White

FIGURE 1.12 Apollo 16 Command and Service module over the Moon. SOURCE: Courtesy of NASA; GPN-2002-000069, available at http://grin.hq.nasa.gov/.

House and NASA believed that the Astronaut Office should "look like America." To ensure that the broadest population possible could apply, the space shuttle physical anthropometric standards were enlarged to allow larger men (95th percentile) and smaller women (5th percentile).

Pilot astronauts would be required to maintain high-performance proficiency in the T-38. Mission specialists were required to qualify as backseat T-38 crew members (Figure 1.15).

At that time, NASA also developed plans for a third type of astronaut, the payload specialist. Payload specialists were not career astronauts; they were to be selected as needed for specific missions (Spacelab, military payloads, commercial satellite deployments, and international payloads), fly once, and then return to their other

FIGURE 1.13 Astronaut Russell L. Schweickart stands on the Lunar Module's "Spider" porch during an EVA on the Apollo 9 mission. SOURCE: Courtesy of NASA/James McDivitt. GPN-2000-001108, available at http://grin.hq.nasa.gov/.

jobs. They would not be trained to operate the RMS or to do EVAs, nor would they fly the T-38, except for one or two "familiarization" rides with their mission commanders.

In 1978-1986, approximately 60 payload specialists were trained at JSC. Flights dedicated to research trained up to four payload specialists for each mission position; one or two primary payload specialists were selected within 6 months of flight by the payload investigator working group in consultation with the Astronaut Office. The Astronaut Office had veto power if it concluded that an astronaut was not "safe" enough to function in the dynamics of the spaceflight environment or as an integral part of the crew.

The Astronaut Corps in 1985 was just large enough to support the flight rate and crew training template. In some cases, mission specialists in particular were rotating from one flight to training for another flight with only a few weeks in between.

In January 1986, the NASA Astronaut Corps had over 100 active members; this was considered barely adequate

FIGURE 1.14 Astronaut Alan Bean speaking to engineers at Grumman, the manufacturer of the Lunar Module, in 1966. After the January 1967 Apollo 1 fire, astronauts became more integrated into the design and development of the spacecraft that they would fly. SOURCE: Courtesy of Northrop Grumman History Center.

FIGURE 1.15 T-38N jets in flight over NASA's Dryden Flight Research Center in California. SOURCE: NASA/Jim Ross; ED07-0222-06, available at http://www.nasa.gov/centers/dryden/multimedia/.

to staff and support a shuttle flight rate of 8-10 missions per year and proposed space station missions. A new selection was planned for 1986 when the *Challenger* accident occurred.

The Challenger Accident in 1986 and Implications for the Astronaut Corps

After *Challenger*, the Space Shuttle program was grounded until return to flight in 1988, and almost all astronauts were assigned to various technical jobs to support the investigation and the recertification of the space shuttle and all its components. The Teachers in Space program largely ended, but payload specialists were still being selected. Anticipating the continuation of the Space Shuttle program and the space station, NASA continued to recruit.

The investigation report made clear recommendations about including astronauts into more support roles in the management structure to ensure mission safety. Some astronauts accepted temporary assignments into field center and Headquarters positions, and others eventually left their flight careers to accept permanent civil service positions.

The Post-Challenger Shuttle Era: 1988-2011

In mid-1993, NASA and the newly formed Russian Space Agency (RSA) signed an agreement to initiate a joint program to dock the space shuttle to the Mir space station in the expectation that RSA would join the International Space Station. The new project also called for U.S. support astronauts in Star City who had to acquire Russian language skills, and this eventually became a training opportunity for future Mir crew selection. The project required more astronauts than had been anticipated in the staffing models, and Russian language was added to the NASA astronaut training curricula. (See Figures 1.16, 1.17, and 1.18.)

The ISS Era

With the development of the International Space Station (an ISS training cycle could take 4 years for each astronaut) came the expectation that astronauts would again be required and expected to support the new vehicle design. The agency was flying four or five shuttle missions each year in addition to ISS expeditions, and the Astronaut Office reached its peak staffing size of nearly 150 in 1999. Mission support was intense and geographically diverse. The planet-wide geographical spread of training strained crew time to support ISS design and safety boards at the various NASA centers. (See Figure 1.18.)

By 2004, however, the *Columbia* accident and the planned end of the Space Shuttle program caused NASA to slow and shrink astronaut candidate recruitments. Eleven were selected in June 2004; these were the first astronaut candidates to be told that they might not fly on a shuttle (all did). An additional nine were selected in June 2009. (See Figures 1.19 and 1.20.)

HISTORY OF THE ORGANIZATIONAL STRUCTURE AND ROLE OF THE ASTRONAUT OFFICE

The Astronaut Office is a part of the Flight Crew Operations Directorate, which also includes Aircraft Operations. This structure has matured over the last 40 years but has varied little.

The Astronaut Office, like many other flight crews in "test and development" organizations, is internally organized to support both the ongoing missions with trained crew (this drives the office size) and mission development and safety. Many of the roles would be considered normal operations in other flight test organizations, such as the Air Force Flight Test Center at Edwards AFB, California, the Navy Flight Test Center at Patuxent River, Maryland, and the commercial Boeing test flight program in Seattle, Washington. Astronauts are required to sign off on multiple steps of the safety process, including the Certification of Flight Readiness (CoFR), before each spaceflight.

Although selected to support the mission manifest, the Astronaut Corps is also used as subject matter experts in the development of future spacecraft and research payloads by providing design expertise and lessons learned. Astronauts selected for specific engineering or scientific expertise (such as materials scientists, astronomers, and

FIGURE 1.16 Russia's Mir Space Station seen against a blue and white backdrop of Earth by the space shuttle *Atlantis* after undocking in June 1995. SOURCE: Courtesy of NASA; available at http://spaceflight.nasa.gov/history/shuttle-mir/multimedia/sts-71-photos/71p-021.htm.

physicians) were assigned to help in the design of payloads or experiments and engineering tests identified for flight.

Astronauts are also responsible for evaluation, testing, and development of new vehicle designs, hardware, and operations (such as the rapid prototyping laboratory). They formally sit on and are voting members of a number of design and safety boards. They are integral to the development of flight procedures, which generally change with each mission.

HISTORY OF NASA GROUND TRAINING FACILITIES AND ALLOCATION FOR SPACEFLIGHT READINESS TRAINING

Astronaut training must cover a wide array of skills, which require a variety of training facilities. The definition, design, and funding of ground-based simulators is the responsibility of an organization in the Mission Operations Directorate (MOD). MOD is also the parent organization of the Mission Control Center (MCC) and its flight controllers. MOD works with the Astronaut Office in the development of simulator and training requirements. Until the late 1980s, MOD was also in FCOD.

Computer-driven simulators were first developed in support of the Mercury, Gemini, and Apollo programs (Figures 1.21 and 1.22). Many of the simulators, including the ones for the Lunar Module and the Lunar Rover,

FIGURE 1.17 Crew members from STS-71, Mir-18, and Mir-19 pose for an in-flight picture. SOURCE: Courtesy of NASA; GPN-2002-000061, available at http://grin.hq.nasa.gov/.

FIGURE 1.18 The International Space Station orbiting Earth as seen from STS-102. SOURCE: Courtesy of NASA, available at http://mix.msfc.nasa.gov/abstracts.php?p=1644.

were primarily mechanical emulators. Others, such as the capsules, were high-fidelity physical environments and could cause instruments and surrounding environments to emulate both normal operations and emergency operations. The training organization strove to keep the trainers under configuration control so that there would be no surprises for the crew when they operated the real vehicle.

The MOD Training Division works closely with FCOD to ensure that simulators and other trainers provide the appropriate fidelity, and both MOD and FCOD are responsible for ensuring that crews pass training tests before flight. The introduction of each new spacecraft and new tasks (such as EVAs and rendezvous and docking) led

FIGURE 1.19 NASA astronaut candidates Christopher Cassidy (*left*), Jose Hernandez (*middle*), and Japan Aerospace Exploration Agency astronaut Naoko Yamazaki (*right*) practice their navigation skills. SOURCE: Courtesy of NASA Johnson Space Center Features; available at http://www.jsc.nasa.gov/jscfeatures/articles/000000255.html.

FIGURE 1.20 NASA astronaut candidates conduct an emergency egress drill during land survival training. SOURCE: Courtesy of NASA Johnson Space Center Features; available at http://www.jsc.nasa.gov/jscfeatures/articles/000000255.html.

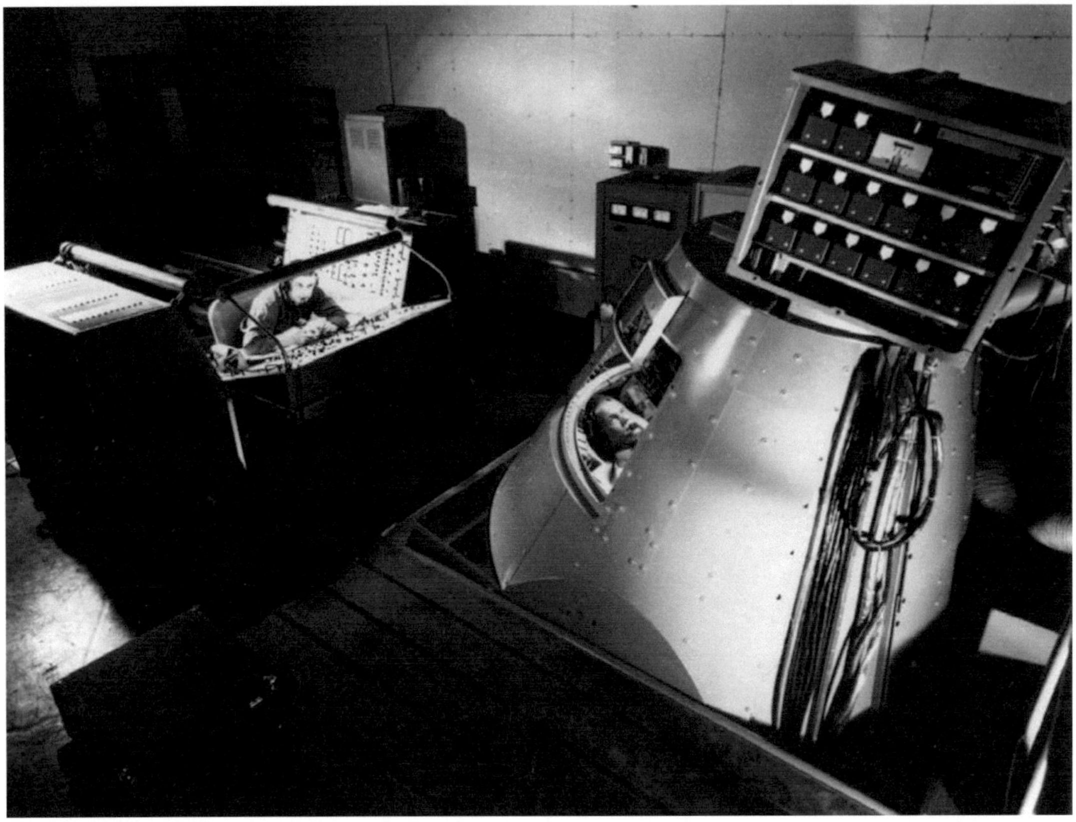

FIGURE 1.21 Astronaut John H. Glenn in the Mercury Procedures Trainer. SOURCE: Courtesy of NASA; GPN-2002-000044, available at http://grin.hq.nasa.gov/.

to the introduction of new training equipment and facilities to support them. Many of the facilities were at the Johnson Space Center; others were at NASA centers across the country (Figure 1.23). Over time, astronaut training facilities were consolidated at JSC, including the Neutral Buoyancy Laboratory (NBL), the large swimming pool used by astronauts to train for EVAs. The shuttle era introduced a large number of facilities and trainers to train astronauts in such diverse tasks as flying and landing the shuttle, operating the RMS, conducting emergency procedures, EVA preparation, operating the Spacelab, and using various shuttle systems.

The ISS era had a major impact on the requirements for training facilities. Each new ISS segment could, in effect, be considered a new spacecraft. Astronauts needed to train for basic ISS tasks, such as stowing equipment, and more advanced tasks, such as operating system and science software and maintaining complex ISS systems. These different requirements resulted in the development of both low-fidelity and high-fidelity ISS simulators. ISS training was also complicated by the introduction of international components, such as laboratory modules, and new, non-U.S. spacecraft, such as the Soyuz. In general, NASA adopted a policy of having low-fidelity mockups of international equipment at JSC and sending astronauts overseas to the international partners who possessed the high-fidelity trainers and equipment (Figure 1.24).

Ground-based simulators for space fight missions are used for 90-95 percent of training. As little as 5 percent of crew training time is spent with SFRT in preparing for a mission, but it is considered by FCOD to be a critical part of crew training and the primary distinction between preparing for an Earth-based mission and one in space.

FIGURE 1.22 The Project Mercury altitude wind tunnel gimbaling rig. SOURCE: Courtesy of NASA/Bill Bowles; GPN-2000-000385, available at http://grin.hq.nasa.gov/.

FIGURE 1.23 Two astronauts practice construction techniques to build a space station in the Neutral Buoyancy Simulator at Marshall Space Center in 1985. In the 1990s, EVA training was consolidated at the Johnson Space Center. SOURCE: Courtesy of NASA/Dennis Keim; GPN-2000-000057, available at http://grin.hq.nasa.gov/.

Apart from SFRT, small environmental additions to flight training include exposure to hypoxia in the JSC high-altitude chamber, a one-time run in the Russian centrifuge at Star City (for medical evaluation after exposure to ballistic re-entry), and survival training (water and land).

SUMMARY

The U.S. astronaut program has evolved over the decades to meet the needs of the new activities initiated by NASA. But it has also adapted to new social and political realities. The program has incorporated non-test pilots, scientists, a much broader demographic base, international participants, and other groups. Changes in training have been driven not only by the introduction of new spacecraft and requirements but by the need to accommodate astronauts whose experience is different from that of test pilots. Now, with the shuttle retired and the ISS having entered its fully operational phase, the agency is undergoing a new, and uncertain, transformation, which will also have implications for the Flight Crew Operations Directorate, the Astronaut Office, and the Astronaut Corps.

FIGURE 1.24 STS-131 preparing to dock with the International Space Station on April 7, 2010. SOURCE: Courtesy of NASA.

2

NASA's Human Spaceflight: The Role and Size of Activities Managed by the Flight Crew Operations Directorate

The committee's task charges it with addressing how the role and size of the activities that are managed by the Johnson Space Center Flight Crew Operations Directorate (FCOD) should change after space shuttle retirement and completion of the assembly of the International Space Station (Figure 2.1). NASA's Flight Crew Operations Directorate and the Astronaut Office are undergoing a major transition as the Space Shuttle program ends and focus shifts to operations on the International Space Station (ISS). The question of the role of FCOD (and its Astronaut Office) and the activities that it executes are directly related to the requirement for astronauts, that is, the size of the Astronaut Corps.[1] Although the shuttle is now retired, new spacecraft will be selected to provide ISS support, and the agency is evaluating possible missions beyond low Earth orbit. This chapter addresses the current size of the Astronaut Corps, ISS training requirements, the requirements for supporting new missions, and how these requirements will affect the size of the Astronaut Corps.

NASA HUMAN SPACEFLIGHT MISSION REQUIREMENTS

NASA's strategic requirements for human spaceflight are dominated by the International Space Station. The agency currently plans to operate the ISS until at least 2020, and this suggests that astronauts must be trained to operate the ISS and its systems safely for at least the next decade while new systems, such as commercial robotic resupply and transport, are incorporated.

NASA conducts extensive training of its astronauts before they are approved to fly on a mission. The current

[1] The size of the Astronaut Office has been the topic of several previous NASA internal studies. A 1993 report by NASA reached several conclusions regarding the NASA Astronaut Corps. The number of astronauts in the corps had increased to 104 in 1992 because of longer, more involved shuttle missions. In response to an audit of flight crew training at Johnson Space Center (JSC), the report found that JSC had appropriately responded to recommendations by increasing the number of training instructors and upgrading the Shuttle Mission Training Facility. The personnel and shuttle facilities were deemed sufficient to meet the current flight rate. A 2003 report by the NASA Office of the Inspector General examined the size and utilization of the Astronaut Corps. The report found that several issues in the astronaut selection process contributed to unnecessarily high costs of astronaut training and that the training was not being used appropriately relative to the expense incurred. The report recommended that NASA establish formal guidelines for the astronaut candidate selection process and determine a more realistic assessment of the necessary size of the Astronaut Corps while documenting any reason for deviating from the assessment. U.S. General Accounting Office, *Results of General Accounting Office Survey of NASA Astronaut Utilization,* NSIAD-93-114R, January 12, 1993, Washington, D.C.; NASA Office of the Inspector General, *Improving Management of the Astronaut Corps,* G-01-035, June 27, 2003, Washington, D.C.

FIGURE 2.1 Sunrise over Earth and the International Space Station. SOURCE: Courtesy of NASA.

Astronaut Office management presented a comprehensive briefing to the committee on its current mission. At a top level, the current Astronaut Office mission statement is to support five tasks:

1. Provide well-trained spaceflight operators to support the NASA flight manifest.
2. Provide ground support personnel for unique tasks required to support the NASA flight manifest.
3. Provide support for new program development. This ranges from development of relatively small payloads and equipment to new spacecraft designs.
4. Provide support for public and educational outreach to our society.
5. Provide for collaboration with other government and private organizations as needed.

Although NASA did not present this list in priority order, the committee believes that the first task, providing well-trained spaceflight operators to support the NASA flight manifest, is clearly the Astronaut Office's top priority, and the remaining tasks are at least in rough priority order. The committee identified a sixth task: Provide operational knowledge and corporate memory of human spaceflight.[2] The committee concluded that the Astronaut Office *already* provides this operational knowledge and corporate memory of human spaceflight. For example, senior astronauts train new astronauts, and astronauts who have finished flying space missions often transition to management positions in the agency and take their knowledge with them. NASA has frequently used astronauts in senior leadership positions where they run major programs and even centers. With the transition to commercial crew operations, the committee concluded that it was important to make this role of the Astronaut Office explicit, not merely implicit.

[2] In Finding 2.1a, the committee has listed NASA's five tasks plus the sixth, added by the committee, in the priority order established by the committee, which places public outreach and education sixth on the list.

FIGURE 2.2 The Jules Verne Automated Transfer Vehicle, or ATV (*left*), and the Japanese H-II Transfer Vehicle, or HTV (*right*), orbiting Earth. SOURCE: Courtesy of NASA.

Although the retirement of the space shuttle has reduced some of the training requirements for NASA astronauts, the operation of the ISS has imposed many complicated new ones, such as the requirement for Russian language proficiency. Astronauts are now required to be familiar not only with the U.S. equipment aboard the ISS but European, Japanese, and Russian station modules and equipment. They are also required to be knowledgeable about the Soyuz spacecraft and the Progress, Japanese HTV, and European ATV robotic resupply spacecraft (Figure 2.2). They must be proficient in using space station software, conducting extravehicular activities, operating the space station's robotic arm, and numerous other tasks. Furthermore, astronauts are no longer trained for focused, limited duration missions with clearly defined skill sets (as they were with the shuttle) but instead are required to have the knowledge and skills to live in space for a long duration, respond to an eventuality that may arise (and fix it there rather than return to Earth), and conduct a large array of science experiments.

Since 2009, there has been considerable debate and disagreement between Congress and the White House about the future direction of the U.S. human spaceflight program. There is no clear plan to send U.S. astronauts beyond low Earth orbit in the foreseeable future, but it remains a possibility, particularly in light of NASA's recent announcement of its intention to develop a Multi-Purpose Crew Vehicle for the follow-on exploration of space.

In many ways, crew roles, tasks, and skills for human exploration are the same as those currently required for the ISS and Soyuz operations: long-duration spaceflight, on-orbit crew stay; ascent and entry; rendezvous and docking; undocking and de-orbit; maintenance and repair; robotics operations; EVA operations; and scientific research and payload operations. For human exploration and operations beyond low Earth orbit, the ISS task and skill set will need to be augmented by training for planetary surface operations, mission-specific operations and landing requirements, and science operations. Although the specific missions have not been approved, the potential mission set includes the Moon, Mars, near-Earth asteroids, and spacecraft servicing.

NASA is also planning to use commercially procured crew transfer services for the ISS, which may or may not involve use of NASA astronauts for operations. However, even if the NASA astronauts are not used to pilot the commercial vehicles, the committee believes that NASA's Astronaut Corps will be involved in various aspects of development and certification of commercial service provider pilots to provide an oversight role and to ensure safety.

CURRENT ASTRONAUT CORPS STAFFING

NASA's Astronaut Corps of active, flight-eligible astronauts is managed by the Astronaut Office, an organization in Johnson Space Center's Flight Crew Operations Directorate. FCOD manages both the Astronaut Office and aircraft operations and training at Ellington Field several miles north of the Johnson Space Center. FCOD managers determine the staffing size of the Astronaut Corps to meet expected U.S. on-orbit segment crew requirements aboard the International Space Station.

NASA has experienced considerable attrition concurrent with the shuttle's final missions and retirement. Although historical attrition during the Apollo-to-shuttle transition is a poor predictor of future turnover (no flight program beyond the ISS that might retain members of the Astronaut Corps after shuttle retirement is being readied), the size of the Astronaut Corps has now reached the level that it had shortly before the shuttle began flying: it has shrunk by more than 50 percent since 2000, and is expected to shrink further (see Figures 2.3 and 2.4).

Crew member physical size further constrains the flexibility to assign astronauts to fly to the ISS at least until an alternative to Soyuz is developed. The Soyuz has constrained NASA astronaut crew assignments since the beginning of the Shuttle-Mir program in 1995 (some astronauts are too large and others are too small, although modifications to the Soyuz have eased these restrictions). Selected and trained astronauts were still able to fly on the space shuttle to support either the ISS or the shuttle. With termination of the Space Shuttle program, these astronauts have no future flight opportunities and have left or are in the process of leaving the Astronaut Office. That has affected the FCOD models for mix of experienced versus new astronauts and the expected attrition rates and has adversely affected career astronauts who were selected expressly for ISS duty when Soyuz size was not a selection criterion.

As of June 2011, the Astronaut Corps had 61 members, and an additional 9 astronaut candidates selected in 2009 were in training. The corps comprises well-trained spaceflight operators who were hired to crew the space shuttle, the ISS, and Soyuz spacecraft. They are able to conduct flight operations, orbital rendezvous and docking, and scientific research in orbit. Astronauts also perform ground support for spaceflight operations.

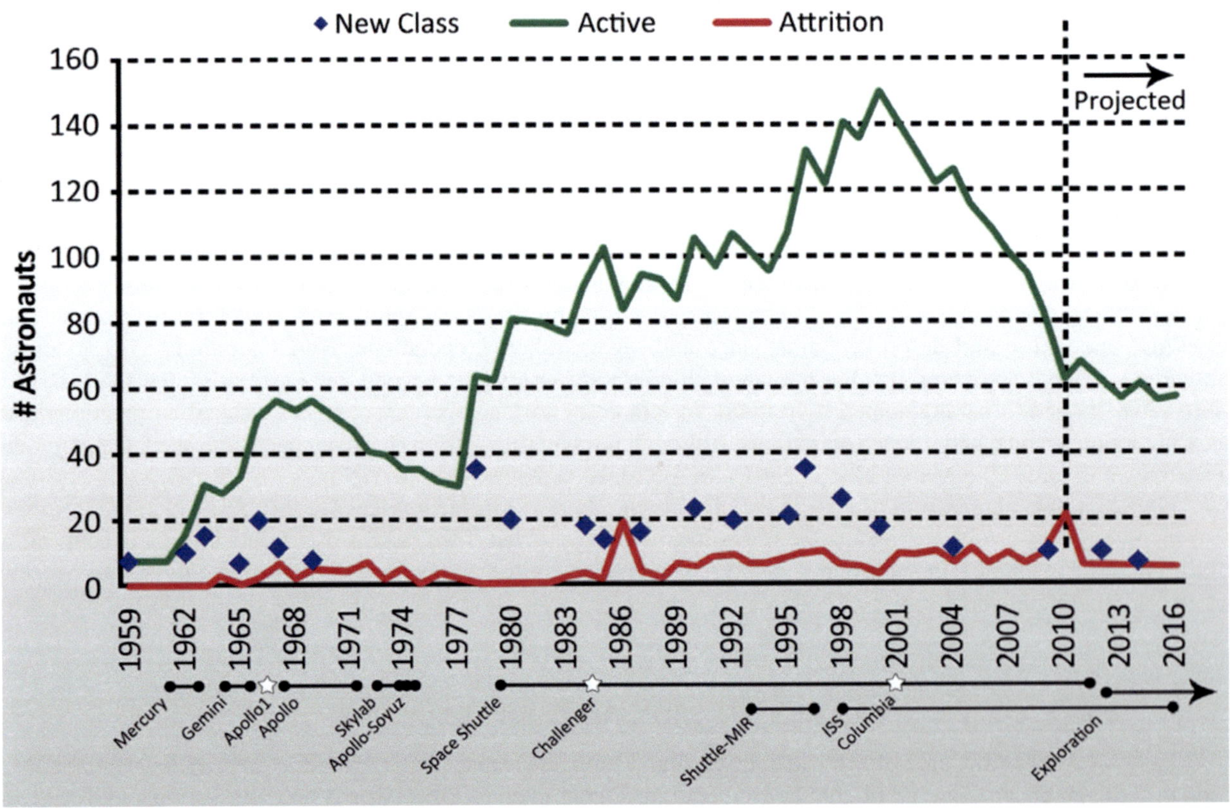

FIGURE 2.3 Historic Astronaut Corps population. SOURCE: NASA Astronaut Office, "Ensuring the Readiness of the Astronaut Corps: A White Paper," NASA Johnson Space Center, Houston, Tex., March 25, 2011.

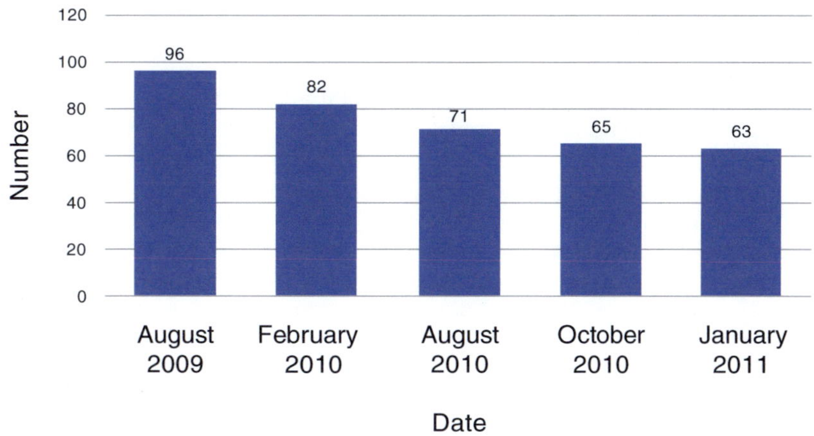

FIGURE 2.4 Recent Astronaut Corps population. SOURCE: NASA Astronaut Office, "Ensuring the Readiness of the Astronaut Corps: A White Paper," NASA Johnson Space Center, Houston, Tex., March 25, 2011.

Astronauts are skilled in evaluation, testing, and development of new vehicle designs, spaceflight hardware, and operations techniques. They also perform educational and outreach duties and collaborate with NASA directorates and other government organizations. The committee notes that the Astronaut Office provides operational knowledge and corporate memory of human spaceflight.

The Astronaut Office is organized into 10 branches and is led by a chief astronaut and a deputy (Figure 2.5).

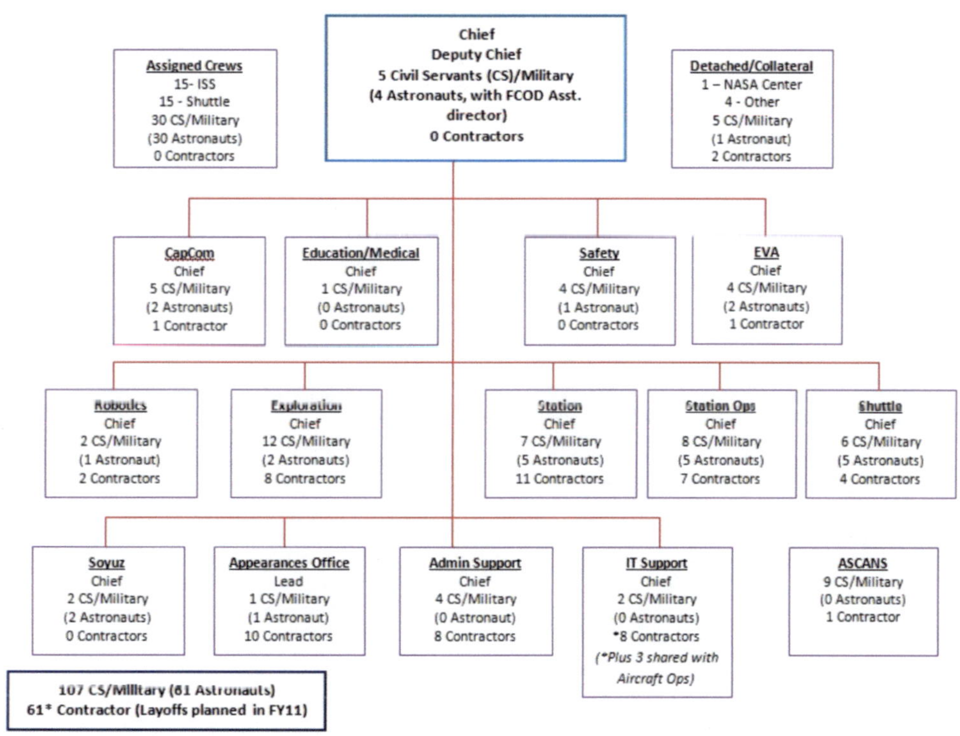

FIGURE 2.5 Astronaut Office organization, as of March 8, 2011. CS, civil servant. SOURCE: NASA Astronaut Office, "Ensuring the Readiness of the Astronaut Corps: A White Paper," NASA Johnson Space Center, Houston, Tex., March 25, 2011.

Nine are directly related to flight support (Capcom, Safety, EVA, Robotics, Exploration, Station, Station Ops, Shuttle, and Soyuz). There are additional support groups that do not have astronauts (Education/Medical, Appearances Office, Admin Support, and IT Support) and other reporting units (Assigned Crews, Detached/Collateral Personnel, and Astronaut Candidates). The Shuttle branch will be dis-established now that the STS-135 mission is completed.

At any given time, a majority of the astronauts in the corps are not assigned to flight. However, they do provide essential support for current and future missions and needs. (See Figures 2.6 and 2.7.)

Some of the support roles are also filled by "management astronauts." These are former astronauts who are

61 Astronauts

- 2 on board ISS
- 11 assigned to ISS missions
- 9 assigned to space shuttle missions
- 6 space shuttle post-flight
- 5 ISS post-flight debrief/reconditioning/public outreach
  - Within 1.5 years of landing
- 1 detached/collateral
  - Director of Operations, Russia
- 9 in leadership roles within the Astronaut Office or the Flight Crew Operations Directorate
- 18 in support roles
  - CAPCOMs in MCC
  - Crew support astronauts
  - Astronaut support personnel
  - Instructor-astronauts

FIGURE 2.6 Current astronaut roles, as of March 8, 2011. SOURCE: NASA Astronaut Office, "Ensuring the Readiness of the Astronaut Corps: A White Paper," NASA Johnson Space Center, Houston, Tex., March 25, 2011.

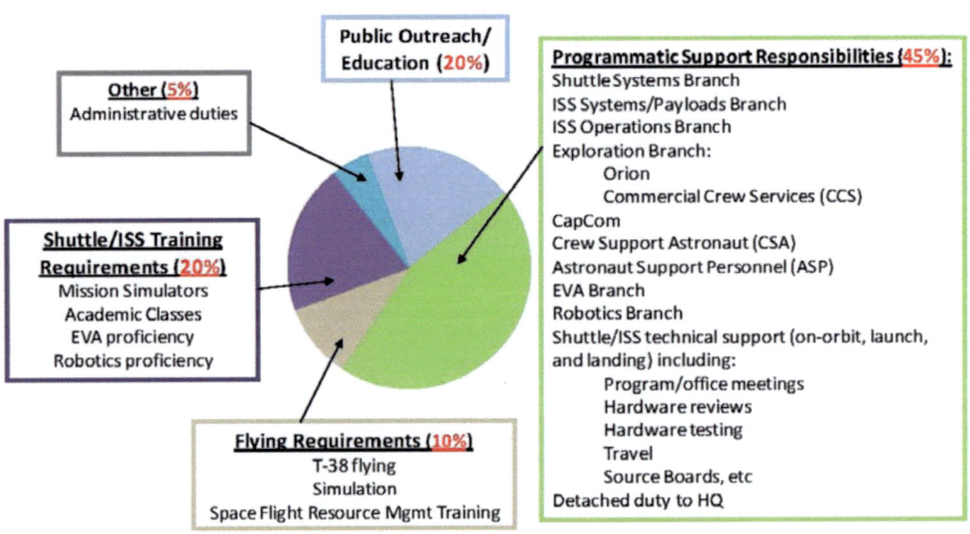

FIGURE 2.7 Unassigned astronaut time distribution. Flying hour requirements are lower for astronauts who are assigned to a mission. SOURCE: NASA Astronaut Office, "Ensuring the Readiness of the Astronaut Corps: A White Paper," NASA Johnson Space Center, Houston, Tex., March 25, 2011.

42 Management Astronauts (Not Eligible for Assignment)

- 17 support the Astronaut Office at the Johnson Space Center (JSC)
    - 4 Astronaut Office branch chiefs
    - 2 CAPCOMs in MCC
    - 1 astronaut support personnel
    - 1 ISS payload lead
    - 5 Exploration/CCS
    - 4 detached/collateral
        - NASA representatives to the U.S. Air Force Space Command and the U.S. Northern Command
        - NASA Glenn Research Center
        - Texas A&M University
        - U.S. Naval Postgraduate School in Monterey, California
- 25 outside the Astronaut Office at NASA
    - 3 Headquarters (including NASA Administrator and the Associate Administrator for Education)
    - 19 JSC (including the JSC Center Director, the JSC Deputy Center Director, the Director of Safety and Mission Assurance, the Director of FCOD)
    - 1 Kennedy Space Center (Director, Kennedy Space Center)
    - 1 Goddard Space Flight Center
    - 1 Ames Research Center

FIGURE 2.8 Current management astronaut roles. SOURCE: NASA Astronaut Office, "Ensuring the Readiness of the Astronaut Corps: A White Paper," NASA Johnson Space Center, Houston, Tex., March 25, 2011.

no longer eligible for flight assignment and who do not use NASA aircraft or other training facilities except as instructors (see Figure 2.8).

NASA furnishes 4-6 astronauts annually to fly on the ISS; each astronaut flies a 6-month expedition. Soyuz transports are used for launch and entry and as an emergency escape vehicle at the station (Figure 2.9).

FIGURE 2.9 The Soyuz TMA-12 prepares to dock at the International Space Station on April 10, 2008. SOURCE: Courtesy of NASA.

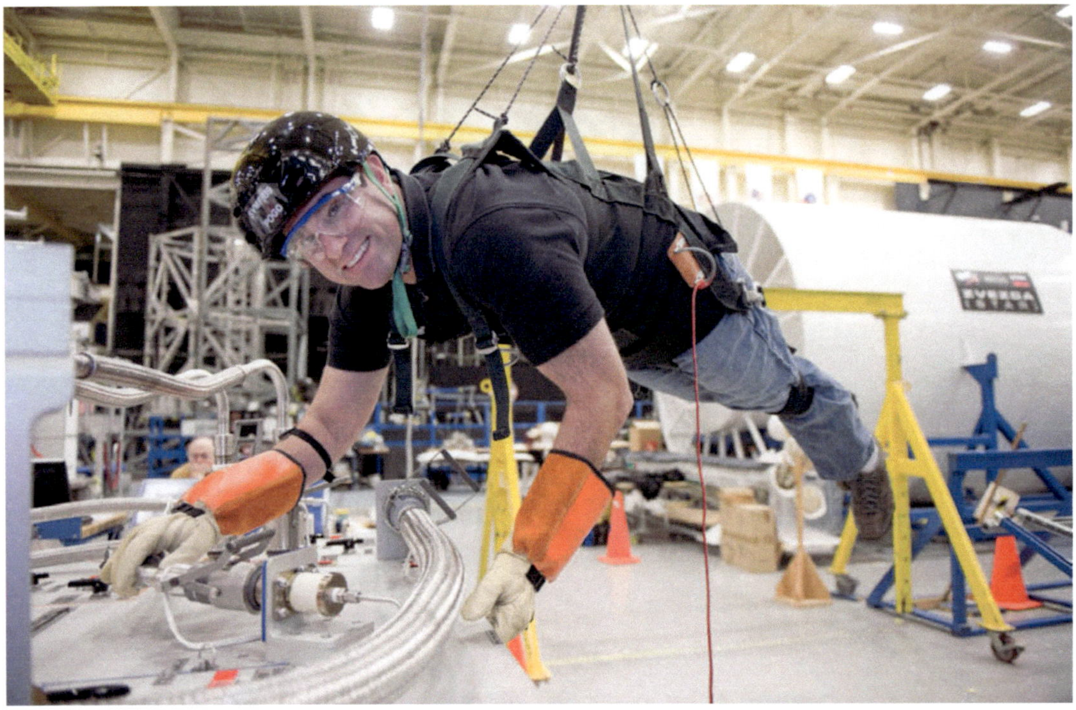

FIGURE 2.10 A NASA astronaut participates in a spacewalk training session in the Partial Gravity Simulator test area in the Space Vehicle Mock-Up Facility at NASA's Johnson Space Center. SOURCE: Courtesy of NASA/Jack Pfaller.

Soyuz training requirements have gained additional prominence with the shuttle's retirement. At present, all ISS crew exchanges occur via Soyuz, and U.S. astronauts must perform Soyuz duties competently as part of the three-person crew. A Russian cosmonaut always serves as the Soyuz commander, and NASA astronauts fill the role of flight engineer (essentially as co-pilot). With the exception of simple familiarization training classes in Houston, astronauts travel to the Gagarin Cosmonaut Training Center (also known as Star City) near Moscow for all Soyuz training. Soyuz classroom, mockup, and simulator training require substantial additional time, as do language training and travel logistics.

In addition to their own expedition training, members of the Astronaut Corps in Houston support ISS operations. They act as crew support for their colleagues in orbit and as capsule communicators in mission control. Astronauts also develop and validate extravehicular activity, robotics, and emergency procedures for use aboard the ISS; act as instructor astronauts; perform hardware fit checks; review training for ISS experiments; monitor cargo orbital transportation services; support international cargo vehicle operations; and take part in the development of NASA and commercial crew exploration vehicles (Figure 2.10).

INTERNATIONAL SPACE STATION TRAINING REQUIREMENTS

It takes a minimum of 3 years to train and fly an astronaut for an ISS mission as opposed to 1 year to train and fly a shuttle astronaut. Variables for flight assignment include suitability for EVA and robotics, Russian language aptitude, and long-duration medical standards. Russian language proficiency is a major training challenge but may be somewhat relaxed when a U.S. vehicle assumes transport and emergency "lifeboat" functions at the ISS. However, intermediate Russian language proficiency will always be necessary for close work aboard the ISS with Russian cosmonauts (Figure 2.11).

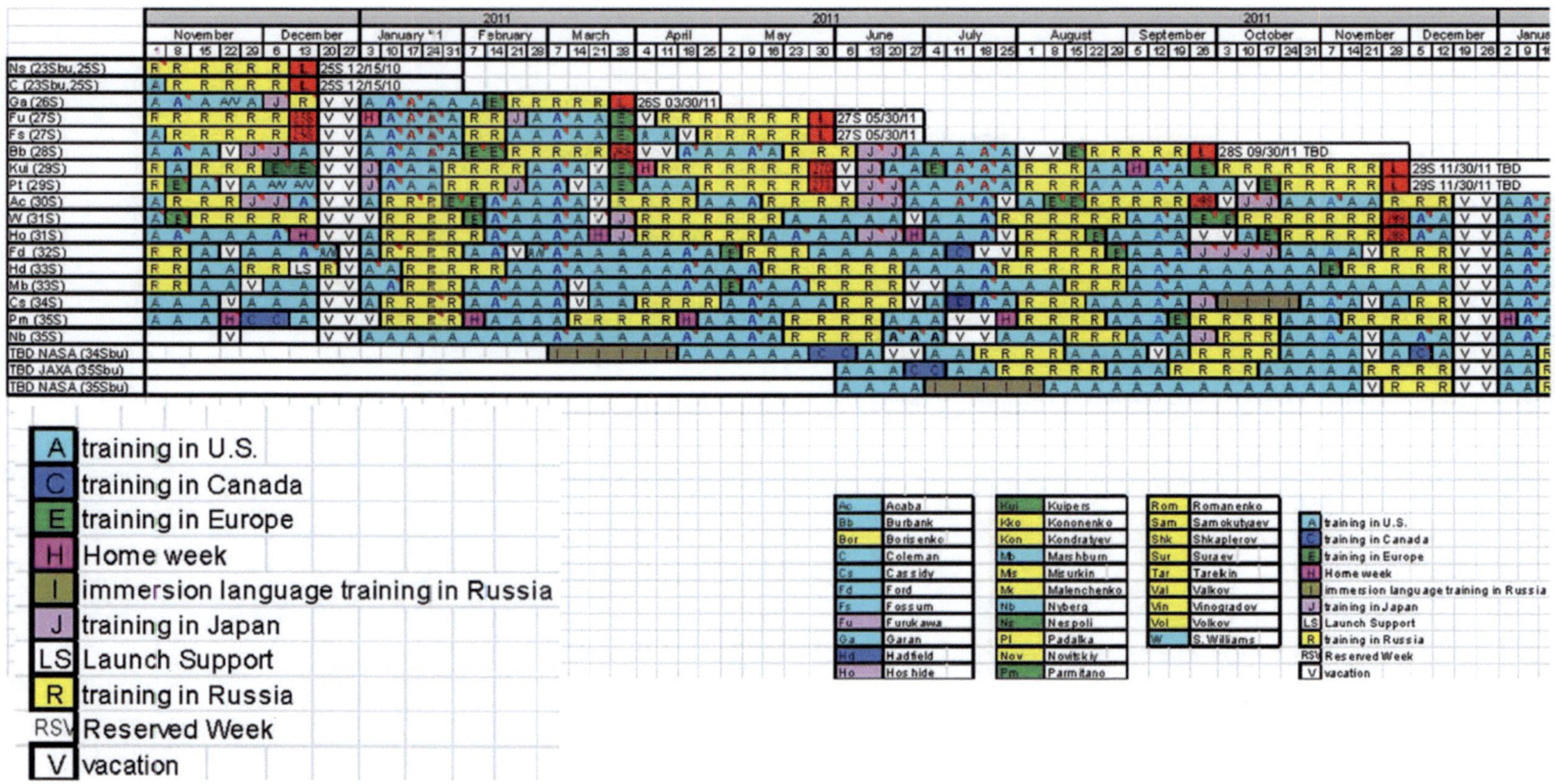

FIGURE 2.11 Typical time slice of ISS international training requirements (using the six-person model and baselined flows as of December 2010). The chart demonstrates the complexity of the training schedule, including the large periods spent overseas and in language training. SOURCE: NASA Astronaut Office, "Ensuring the Readiness of the Astronaut Corps: A White Paper," NASA Johnson Space Center, Houston, Tex., March 25, 2011.

The training of U.S. and international crew for the ISS encompasses the following:

1. Extravehicular Activities.
2. Canadian Robotic Arm operations.
3. Habitability.
4. ISS Systems.
5. U.S. Destiny module.
6. ESA Columbus module.
7. Japanese Kibo module.
8. JAXA visiting cargo vehicle (HTV).
9. ESA visiting cargo vehicle (ATV).
10. Russian visiting vehicles: Soyuz and Progress.
11. Survival Training (Soyuz-centered: water and land).
12. Experiment Research training.
13. Language Training (Russian).
14. Spaceflight readiness training (SFRT).

Those ISS training objectives are set by the International Space Station program and its international partners and are applied to all ISS participants. (Note that the training objectives are for the professional Astronaut Corps, not spaceflight participants who visit the ISS for only a few days.) ISS requirements have increased the number of U.S. and partner crew training hours, especially for the Soyuz "Flight Engineer-1, FE-1" position (left seat).

A summary of ISS training requirements is shown in Figure 2.12.

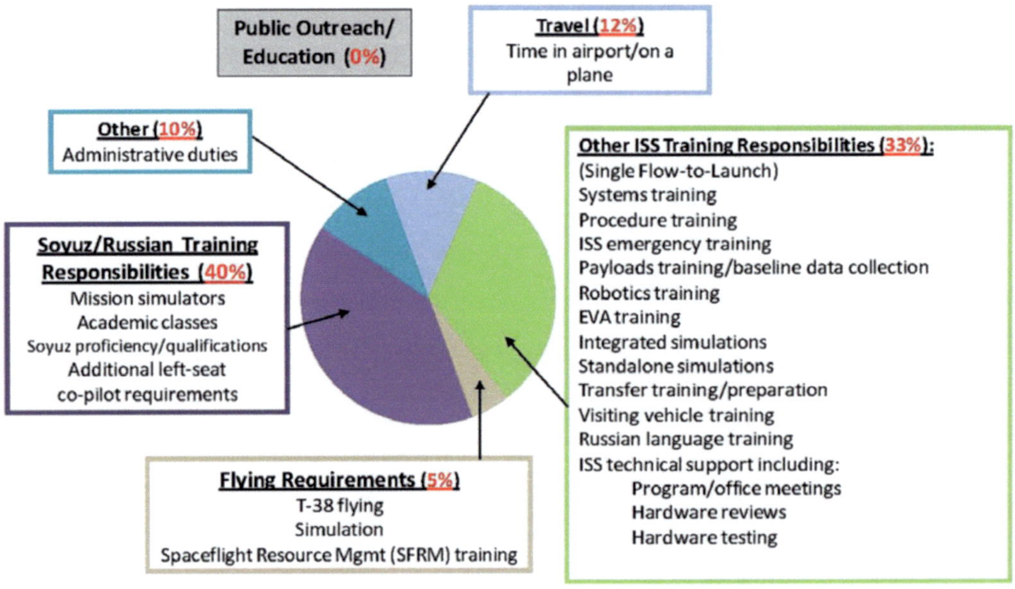

FIGURE 2.12 In the course of a 2.5-year training period, a Soyuz left seat Flight Engineer's (FE-1) work time is divided among Soyuz/Russian training, other ISS responsibilities, travel, administrative duties, and flying requirements (generally known as Spaceflight Resource Management [SFRM] or Spaceflight Readiness Training [SFRT]). Fewer flying hours are required because this is the breakdown for an astronaut who is assigned to a mission and who therefore has less time for T-38N flights. Note that 12 percent of ISS astronaut training time is consumed by travel. SOURCE: NASA Astronaut Office, "Ensuring the Readiness of the Astronaut Corps: A White Paper," NASA Johnson Space Center, Houston, Tex., March 25, 2011.

The requirements for preparing and conducting spaceflight include learning and maintaining proficiency in spaceflight operations and emergency response, as well as mission-specific training. To ensure that crew requirements are addressed, the Astronaut Office requires that astronauts participate in many safety reviews (a requirement since the *Challenger* and *Columbia* accidents); participate in equipment and software design for future systems, projects, and research; evaluate human factors for new vehicles and crew equipment; and certify new crew procedures in high-fidelity simulators that provide appropriate realism. In addition, experienced crew members are also called on to support agencywide panels and task forces involved in strategic planning, to support the Mission Control Center during missions, and to help to communicate the wider education message to the public as dictated in the 1958 Space Act.

EVOLUTION TO POST-SHUTTLE OPERATIONS

As NASA retires the space shuttle and begins a second decade of International Space Station operations, the skill mix that it requires of its astronauts is changing. During the shuttle era, NASA hired two specialized kinds of members of the Astronaut Corps: shuttle pilots and shuttle mission specialists. Pilots monitored and flew the shuttle during launch and landing and performed orbital maneuvering and most major vehicle operations. Mission commanders came from the pilot ranks, moving to the left seat after one or two flights as pilots. Mission specialists handled flight engineer duties, robotic operations, EVA, and major scientific operations.

In the ISS era, those specializations have disappeared, and all astronauts must be capable of a variety of duties on long-duration expeditions: EVA, robotics, science, vehicle operation, repair and maintenance, emergency response, and transport vehicle operations (Figure 2.13). Notably, ISS commanders may come from non-piloting backgrounds—and several ISS commanders were not pilots when they entered the Astronaut Corps. Commanders of future transport vehicles (as on the Russian Soyuz) will probably be required to have professional flying expe-

FIGURE 2.13 U.S. astronauts participate in an extravehicular activity above New Zealand. Conducting EVAs is one of several tasks for which astronauts train. SOURCE: Courtesy of NASA.

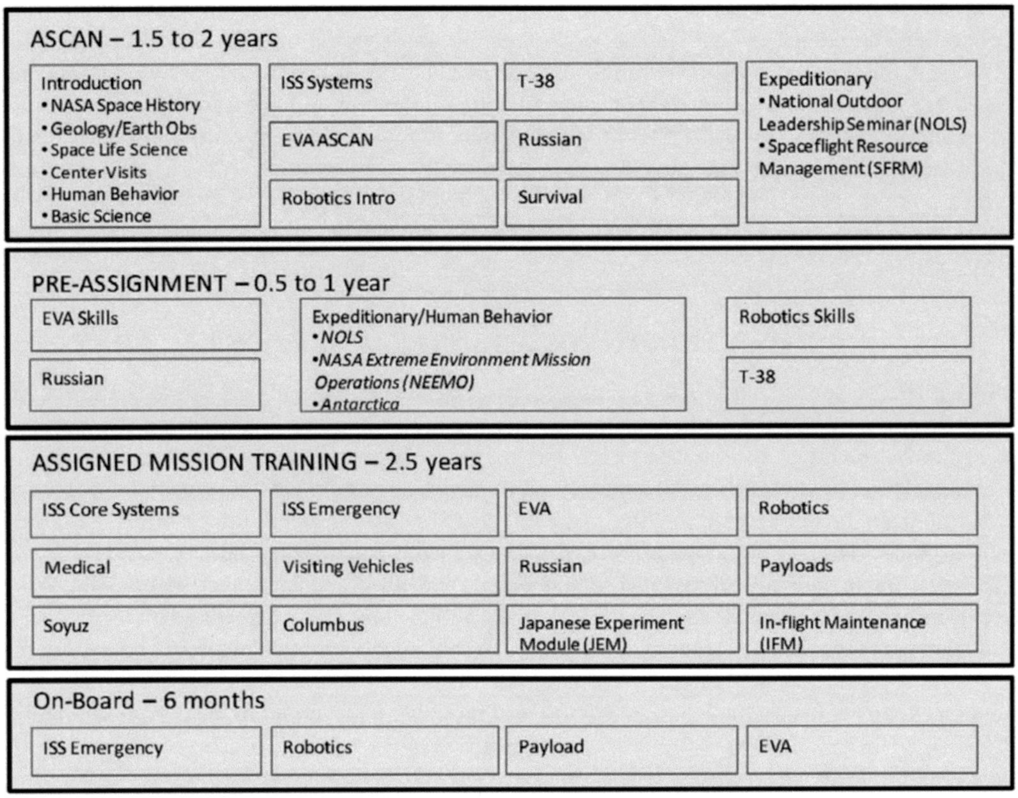

FIGURE 2.14 Summary of International Space Station training—astronaut candidate training through flight training. SOURCE: NASA Astronaut Office, "Ensuring the Readiness of the Astronaut Corps: A White Paper," NASA Johnson Space Center, Houston, Tex., March 25, 2011.

rience or other equivalent experience, such as spaceflight. In the past, the Russians have required either military piloting experience or spaceflight experience for all their Soyuz commanders.

The new roles are reflected in current selection criteria for astronaut candidates. Potential astronaut hires undergo a 2-week interview and medical testing session (versus 1 week for shuttle hires). Candidates must meet long-duration flight physical standards. All potential members of the Astronaut Corps must show aptitude for robotics, science, EVA tasks, and Russian language.

Candidates must also meet the physical size constraints for the Soyuz spacecraft. During the ISS program (and earlier during NASA involvement in the Russian Mir program), a number of astronauts were disqualified for ISS expeditions because they did not meet the Soyuz vehicle seat-size restrictions. Future NASA and commercial vehicles may accommodate a broader size range of astronauts.

ASCAN Training Flow

Astronaut candidates are selected from various professions, including engineering, science, education, and aviation. According to the NASA Astronaut Office, ISS training requires that each crew member have a broad and extensive skill base because there is always some uncertainty as to what tasks will be required. The skills can include payloads, robotics, EVA, and in-flight maintenance. The majority of training is skills-based; however, some critical operations still have a task-based element, such as emergency response. Astronauts from NASA, CSA, ESA, and JAXA train to equivalent levels on U.S. systems, whereas Russian Space Agency cosmonauts are only

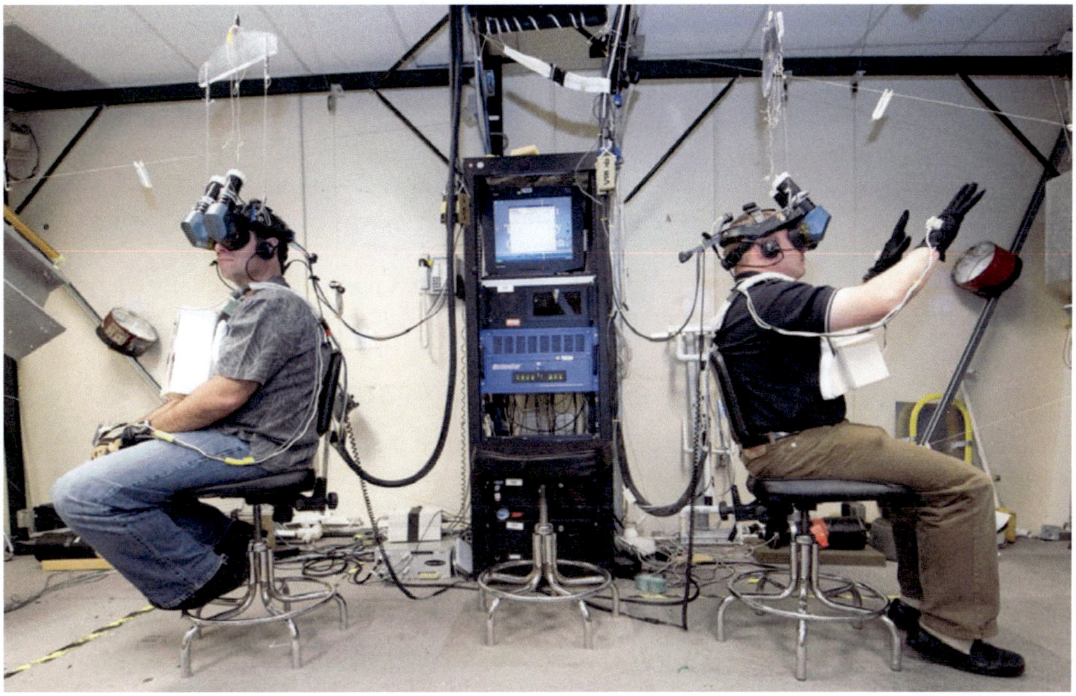

FIGURE 2.15 NASA astronauts train in the virtual reality simulator at the Space Vehicle Mockup Facility at NASA's Johnson Space Center. SOURCE: Courtesy of NASA/Jack Pfaller.

minimally proficient on U.S. systems. All are equally trained in emergency response. Astronauts can expect to do 30-40 percent of their training on international components from Russia, Japan, Canada, and Europe.

International agreements are required for training flows and are led by NASA. Training flows are in four main categories: ASCAN or Basic Training, Pre-Assignment Training, Assigned Crew Training, and On-Orbit Training, all together lasting 5 to 6 years, according to the NASA Astronaut Office. Training is not provided between flight assignments. Astronauts are evaluated before beginning a second flight assignment to test for knowledge and skill retention, after which appropriate refresher courses are given.

ASCAN training comprises four required flows: Inexperienced Operator for ISS Systems, EVA, Generic Robotics Training (GRT), and Russian language proficiency.

Included in the 1.5 to 2 years of ASCAN training are various assessments and evaluations (Figure 2.14). During Pre-Assignment Training, crew members undergo an assessment of aptitude. There are four more pre-assignment training requirements that require 6 months to 1 year to complete: a Russian language intermediate-level proficiency; the U.S. operational segment of EVA; expeditionary training; and robotics, if available.

For Assigned Crew Training, two training flows are available and are based on the astronaut's experience; they generally take 2.5 years to complete. Finally, On-Board Training is given once in orbit and can include a variety of instruction from proficiency and refreshers to unexpected crew tasks. The station training lead on the training team will ultimately decide the training on the basis of crew experience and input from Mission Control Center.[3]

[3] If it appears that a crew member needs some proficiency (such as EVA or remote manipulator or docking) or may even be in the best position to train on something new (such as a newly arrived experiment), that is decided on the ground and then discussed with the commander and the rest of the crew. It is important to note that "training" includes three groups: the crew, the mission control team, and the "training team." The training team is a professional group that acts independently to train the crew and assess their performance, as occurs in other hazardous work, including that of airline and military pilots and submariners. The training function is managed by NASA but is often contracted out to a vendor, such as United Space Alliance.

Astronaut training takes place in a variety of locations. ASCAN training includes classroom activity, site visits, field training, and time in the Space Station Mockup Facility, the Space Station Training Facility, the Dynamic Skills trainer, the Neutral Buoyancy Laboratory, and the Space Vehicle Mockup Facility (Figure 2.15). A few locations are added for the Pre-Flight Assessment, such as the Canadian Space Agency classroom, the NOAA Aquarius Habitat, and remote outdoor locations.

Once a member of the Astronaut Corps has reached ISS training, he or she will spend several weeks at various international training facilities in addition to the U.S. training facilities. For example, a Soyuz right seat Flight Engineer-2 astronaut can expect to spend 49 weeks at U.S. facilities, 2 weeks in Europe, 31 weeks in Russia, 7 weeks in Japan, and 2 weeks in Canada. (The Soyuz Flight Engineer-1 position is essentially equivalent to co-pilot. It is more demanding and requires more training than the Soyuz Flight Engineer-2 position.)

CREW TRAINING IN THE POST-SHUTTLE ERA

Recent developments in commercial spaceflight will require that astronauts learn to operate another new vehicle—possibly more than one—and probably be involved in the development of its displays, habitability, and human factors. Astronaut support and training requirements for these vehicles are currently unknown.

In addition, the agency is embarking on an extensive technology development review for exploration (involving 14 roadmaps). This could result in increased investment in new spacesuit technology, new vehicle designs, surface habitats, and crew-operated rovers. All those technology projects will require corporate knowledge transfer and human factor requirements support from the Astronaut Office.[4]

Crews train for longer periods than ever before. Maximum training time for shuttle/Spacelab was about 18 months. ISS training flows can exceed 2.5 years. The Astronaut Office has worked with the international partners to optimize the training hours and length of training as far as possible without adversely affecting mission success and safety. These requirements are captured in formal agreements at the NASA Headquarters level. Before the ISS, assigned crew could also support other Astronaut Office activities until launch minus 6 months (such as design reviews and safety reviews). That support is no longer possible given the ISS international travel requirements, so the Astronaut Office's ability to support other required flight crew activities is limited further.

With the retirement of the space shuttle, training requirements for crew assigned to the ISS have increased because of the additional travel and language requirements that are placed on the astronauts. On the basis of the committee's assessment of the historically necessary functions of the Astronaut Office and the Astronaut Corps and the future projects that they will probably be asked to undertake, the number of astronauts now available to ensure continuing ground-based operational support and to staff ISS/Soyuz crews appears to be less than required.

The support required for commercial crew vehicles and the new exploration technology development remains undefined. The difficulty of replacing an STS-133 astronaut who sustained a serious injury in a bicycle accident is an indication that the Astronaut Corps is severely strained. The only qualified replacement available was the chief of the Astronaut Office EVA branch. That substitution had an adverse effect when the Astronaut Office was being asked concurrently to participate in a significant restructuring of the agency and to support the continuous operation of the International Space Station. The incident highlights that the Astronaut Corps is reaching a point where it lacks sufficient margin to deal with unexpected personnel situations.

FUTURE STAFFING

The Flight Crew Operations Directorate forecasts the required size of the Astronaut Corps and the need for new hires by using an analytical model that incorporates program requirements, assignment constraints, selection rates, and attrition. The model was incorporated in approximately 2004 and constituted an attempt to apply a more rigorous standard than previous efforts in determining the number of members of the Astronaut Corps required. FCOD

[4] This is the subject of a National Research Council study, conducted by the Aeronautics and Space Engineering Board at the request of NASA's Office of the Chief Technologist, which was under way at the time of the writing of this report. The report of the Committee on the NASA Technology Roadmap is expected in early 2012.

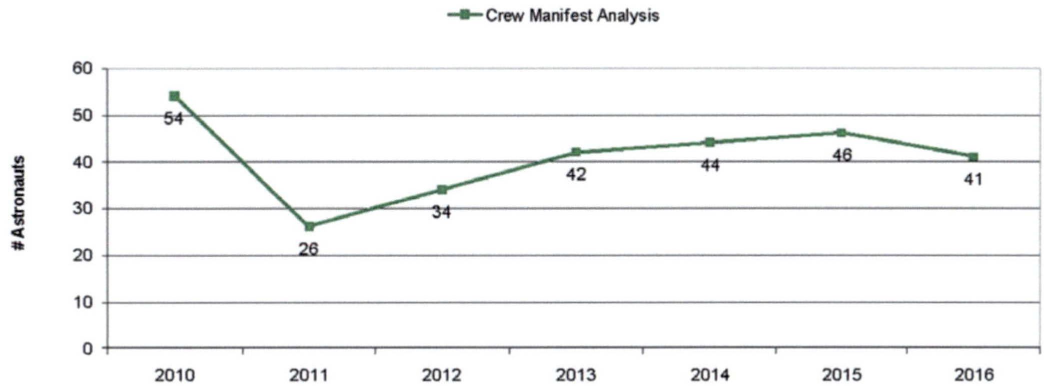

FIGURE 2.16 Crew manifest analysis. NOTE: Data were generated for last Planning, Programming, Budgeting, and Execution in February 2010; assumes ISS ends in 2020 and two space shuttle flights in fiscal year 2010 (STS-133 and STS-134) with additional rescue flight (STS-135) trained and ready to support STS-134 if needed. SOURCE: NASA Astronaut Office, "Ensuring the Readiness of the Astronaut Corps: A White Paper," NASA Johnson Space Center, Houston, Tex., March 25, 2011.

performs the analysis annually as part of its budgeting exercise. The model output is termed the Minimum Manifest Requirement and is calculated from spaceflight program requirements and the 5-year assignment rotation plan.

In the first step, the model sums the number of astronauts in post-flight reconditioning, the number on orbit, and the number of astronauts projected to be needed in space over the next 5 years. The output is the crew manifest analysis, showing the number of astronauts needed by year (Figure 2.16). Note that the model is driven by the manifest requirements, not by the other tasks that astronauts perform, such as programmatic support responsibilities.

To account for astronauts who are unavailable for assignment and to account for constraints (such as desired crew skills mix, temporary medical re-qualifications for flight, and required experience mix), FCOD increases the crew manifest analysis by a factor of 25 percent. The adjusted output is called the Minimum Manifest Requirement (MMR) (Figures 2.17, 2.18, and 2.19).

In 2008 and 2009, the margin applied by FCOD to the model output was 50 percent. In 2010 the margin was decreased to 25 percent, apparently because of budget pressures. To match Astronaut Corps staffing to the MMR, FCOD subtracts annual attrition based on historical averages to produce the astronaut class selections required to bring corps size back to the MMR (Figure 2.18).

Although the MMR analysis was biased in 2011 by the delay in retirement of the space shuttle and by a surplus of space shuttle astronauts relative to the number needed for ISS-only operations, FCOD has confidence that its analytical model and correction factor accurately predict the required Astronaut Corps size.

An important uncertainty is medical: long-duration flight criteria are stringent, and a medical condition

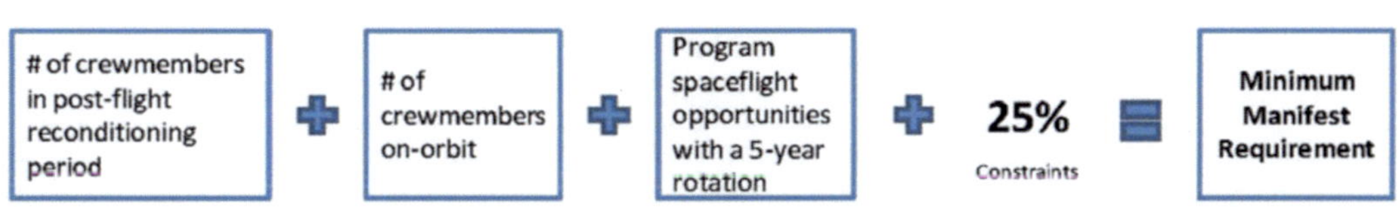

FIGURE 2.17 Minimum Manifest Requirement formula. SOURCE: NASA Astronaut Office, "Ensuring the Readiness of the Astronaut Corps: A White Paper," NASA Johnson Space Center, Houston, Tex., March 25, 2011.

FIGURE 2.18 The crew manifest analysis with the Minimum Manifest Requirement. NOTE: Data were generated for last Planning, Programming, Budgeting, and Execution in February 2010; assumes ISS ends in 2020 and two space shuttle flights in fiscal year 2010 (STS-133 and STS-134) with additional rescue flight (STS-135) trained and ready to support STS-134 if needed; Minimum Manifest Requirement accounts for medical, mission-required skills, attrition, detached/collateral assignments, etc. SOURCE: NASA Astronaut Office, "Ensuring the Readiness of the Astronaut Corps: A White Paper," NASA Johnson Space Center, Houston, Tex., March 25, 2011.

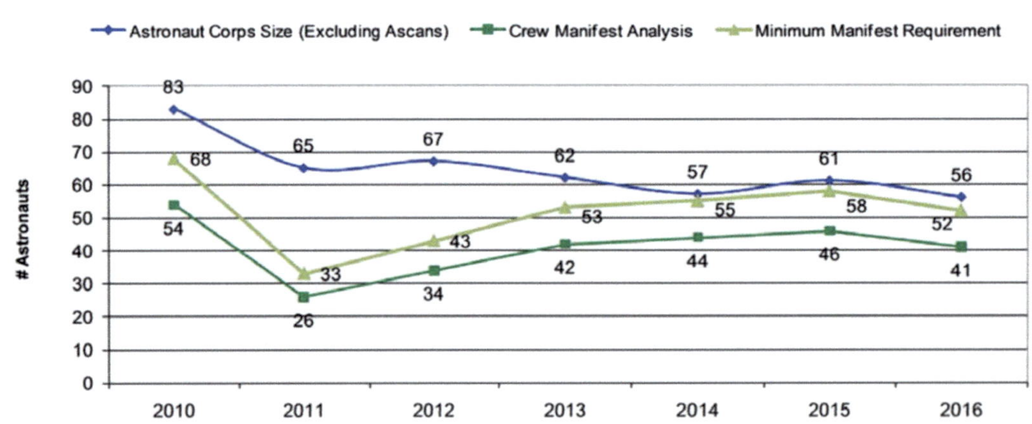

FIGURE 2.19 Crew Manifest Analysis, Minimum Manifest Requirement, and Astronaut Corps size. NOTE: Data were generated for last Planning, Programming, Budgeting, and Execution in February 2010; assumes ISS ends in 2020 and two space shuttle flights in fiscal year 2010 (STS-133 and STS-134) with additional rescue flight (STS-135) trained and ready to support STS-134 if needed; Minimum Manifest Requirement accounts for medical, mission-required skills, attrition, detached/collateral assignments, etc. SOURCE: NASA Astronaut Office, "Ensuring the Readiness of the Astronaut Corps: A White Paper," NASA Johnson Space Center, Houston, Tex., March 25, 2011.

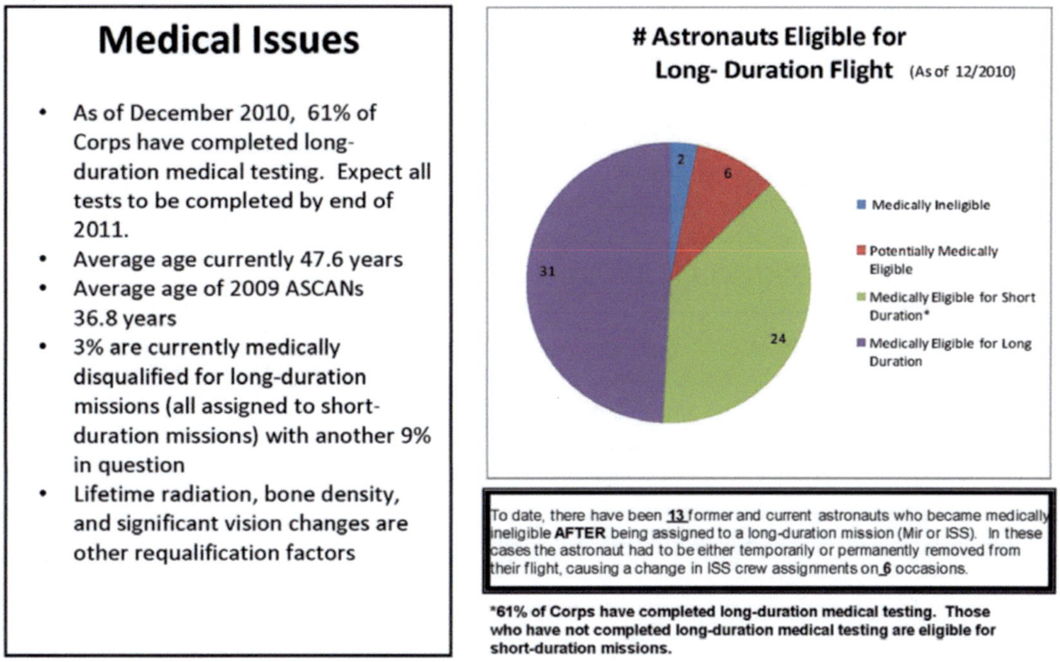

FIGURE 2.20 Medical assignment constraints. The "significant vision changes" are classified as a condition called papilledema, a swelling of the optic disk. This is a relatively new phenomenon in U.S. astronauts and is not fully understood. SOURCE: NASA Astronaut Office, "Ensuring the Readiness of the Astronaut Corps: A White Paper," NASA Johnson Space Center, Houston, Tex., March 25, 2011.

requiring evacuation to Earth is one of the leading risk factors for terminating an ISS expedition early. Thirteen astronauts have become medically ineligible after being assigned to a long-duration expedition but before they actually flew, and this demonstrates the uncertainties that the Astronaut Office must consider when developing the Minimum Manifest Requirement. Because of a variety of medical conditions—including vision problems, bone loss, physical injuries due to EVA, and radiation exposure—some returning astronauts cannot re-qualify for ISS missions (Figures 2.20 and 2.21).

For example, in January 2011, the Astronaut Office needed to assign two members of the Astronaut Corps to meet Expeditions 37/38 and 38/39 requirements. But of 63 Astronaut Corps members available at the time, only 6 were available for the two positions, and not all 6 could meet the more demanding Soyuz left seat Flight Engineer-1 standards.[5] The Astronaut Corps must have reasonable personnel depth to ensure that it can meet on-orbit manning requirements and contingencies with trained astronauts.

Complicating the staffing analysis are such evolving factors as the advent of commercial crew flights to the ISS, a reduction of expedition length to 4 months from the current 6, and a failure to hire a new candidate class in 2012. In addition, although FCOD has calculated attrition on the basis of historical averages, the transition from combined space shuttle and ISS operations to primarily long-duration ISS flights may affect attrition in unknown ways. All those factors could lead to crew shortages within 5 years. Given the lead time for training qualified astronauts, the effects would not be rectified before 2020.

The Astronaut Office projects that it will need to hire 15 new ASCANS within the next 5 years to meet attri-

[5] The Soyuz Flight Engineer-1 position has greater responsibilities and requires more training than the Soyuz Flight Engineer-2 position.

- Radiation exposure limitations are a consideration and vary based on level of solar activity and gender of crewmember
- 7 of 15 crewmembers examined had papilledema postflight (short- and long-duration), with some lingering substantial effects on vision
  - Not always able to re-qualify for subsequent flights
- 5 crewmembers in the last 1.5 yr have had shoulder surgeries attributed to working in the spacesuit
  - 26 injuries in the last year have received rehabilitation services for elbow or shoulder injuries
- Bone loss during spaceflight
  - Largest effects seen in pelvis, trochanter, and femoral neck
  - Recovery may take as long as 3 years for a 6-month mission

Projected Number of Safe Days on ISS Due to Radiation Exposure		
2009 ASCANs	Solar Maximum Activity	Solar Minimum Activity
Females	600 days	409 days
Males	1125 days	783 days

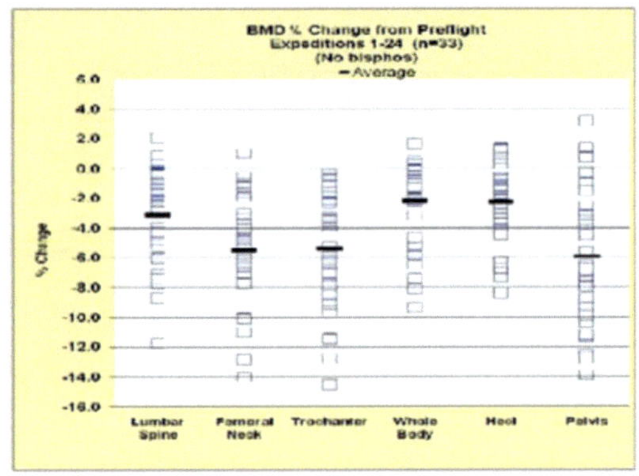

FIGURE 2.21 Medical re-assignment constraints. The projected number of safe days on the ISS due to radiation exposure is affected by the solar cycle. At solar maximum, flares and coronal mass ejections are more common. However, at solar minimum, the solar wind is weaker, and this allows more galactic cosmic rays to enter the solar system and thus reduces the number of safe days on the ISS. SOURCE: NASA Astronaut Office, "Ensuring the Readiness of the Astronaut Corps: A White Paper," NASA Johnson Space Center, Houston, Tex., March 25, 2011.

tion and staffing requirements. Selections are anticipated in 2012 and 2014, on the basis of the February 2010 program estimate.[6] Hiring beyond 2014 will depend on whether the ISS operates beyond 2020 and on progress in future programs. FCOD managers have informed the committee that a steady infusion of small numbers of new astronaut candidates is desirable to introduce new talent, meet demand for specialized skills, and reduce the average age of the corps (the average age of the active astronauts in the corps was 47.6 years as of March 2011).

The ISS flight manifest analysis leading to projected corps size requirements must be considered a minimum. Real-world constraints—such as the need for a mission-specific skill mix, medical disqualification, unwillingness to volunteer for another long-duration expedition, and the need to match veterans with inexperienced astronauts on a given crew—dictate that corps size requirements must always exceed the manifest minimum. That is combined with an uncertain future attrition rate as the agency conducts more longer-duration ISS missions as opposed to shorter-duration shuttle missions. The committee has concluded that the current 25 percent margin will not be sufficient in the coming decade. Given the current difficulties with finding suitable ISS candidates and the likelihood of future commitments to human spaceflight development, the committee's view is that the margin should be increased. Increasing the margin acknowledges that ISS crews require astronauts who have primary skills in engineering and piloting and those who are skilled in the scientific research disciplines corresponding to ISS onboard experiments.

[6] The astronaut selection process takes about 18 months.

POTENTIAL IMPACTS ON REQUIREMENTS FOR THE ASTRONAUT CORPS AND FACILITIES IN THE FUTURE

As the Astronaut Office considers its future staffing needs, hiring frequency, and training requirements for the next decade, its managers must take note of possible responsibilities beyond its primary mission of furnishing astronauts for the International Space Station. Such possible auxiliary missions include:

1. Providing crews for commercial crew transport vehicles on NASA missions.
2. Assisting in the training of commercial astronauts for those vehicles.
3. Providing technical, safety, and operational advice to commercial crew transport developers and the Federal Aviation Administration (FAA).
4. Providing technical liaison and advice to NASA's future vehicle development efforts (for example, heavy lift boosters and spacecraft for beyond Earth orbit travel).
5. Providing crews for testing future NASA vehicles.

Commercial Crew Transport to the ISS

NASA is assisting several commercial companies with funds and expertise in their development of spacecraft to carry cargo, and eventually astronauts, to the International Space Station. The two funded cargo providers are SpaceX and Orbital Sciences Corporation, which are developing the Dragon and Cygnus vehicles, respectively. Because NASA has no government spacecraft that can perform the ISS crew transport mission, the agency anticipates that the Dragon and other spacecraft in development will provide a means to transport NASA astronauts to their ISS assignments and to emergency lifeboat craft at the ISS. Optimistically, such commercial spacecraft may replace the Soyuz in the NASA crew transport role by 2015.

In April 2011, NASA awarded additional development funding to commercial spacecraft developers Blue Origin, Boeing Corporation, SpaceX, and Sierra Nevada Corporation. The funding is intended to enable the Boeing, SpaceX, and Sierra Nevada Corporation (but not Blue Origin) spacecraft to reach a preliminary design review milestone.

Commercial Vehicle Designs

Vehicle designs proposed to date by commercial developers include two types of re-entry shape design: winged or lifting body craft and ballistic capsules. Like the space shuttle, winged or lifting body vehicles develop lift in the atmosphere from the shape of the fuselage or a blend of fuselage and wing shape. Such vehicles are designed to return to runways for landing, and although modern flight control computers are capable of landing such a craft autonomously (for example, the Air Force's X-37B), NASA in the past has always included a pilot "in the loop" to minimize risk during the landing phase. At the moment, only Sierra Nevada's Dream Chaser is a lifting body vehicle. Dream Chaser is being designed for "autoland" capability that would not require a pilot. As in space shuttle pilot training, the safe piloted landing of a winged or lifting body spacecraft requires a high degree of skill and experience. The flying pilot would train for the task and maintain proficiency, as in the Space Shuttle program, through a combination of high-performance aircraft training and simulator training.

Commercial firms have proposed ballistic capsules to serve as ISS crew transports. As in the Mercury, Gemini, Apollo, and Soyuz programs, such vehicles generate a small amount of lift during hypersonic re-entry, enabling some trajectory and aim point adjustment. The blunt body shape creates drag for efficient deceleration from orbital velocity, followed by use of parachutes for terminal deceleration and safe touchdown on land or sea. Although the pilot in Mercury, Gemini, and Apollo could monitor and manually "fly" some portion of the re-entry, in normal operations the spacecraft descends under autopilot control. Ballistic capsules should inherently require much less training in manual flying than winged or lifting body spacecraft.

COMMERCIAL CREW OPERATIONS MODELS

The companies competing to provide commercial crew services to NASA have not yet decided with NASA on the appropriate operations model for transporting crew to the International Space Station. Several models have an influence on NASA astronaut training requirements.

Taxi Model

In the taxi model, commercial astronauts aboard a private spacecraft would carry NASA astronauts to the ISS under government contract. Because the ISS program requires the "taxi" to stay at the ISS for 6 months to provide emergency rescue and safe haven capability, the commercial astronauts would return to Earth aboard the previous rescue or safe haven vehicle. Training requirements would be less stringent, but NASA astronauts would still have to be trained for emergency descent and safe haven operations in case the taxi were required when the commercial astronauts were not onboard the ISS. Although the commercial pilots and crew would be thoroughly trained in all aspects of spacecraft operations, NASA astronauts may still be sought to provide technical, safety, and operational advice to commercial crew transport developers. (A variation of this approach is the harbor pilot model wherein a company pilot would primarily operate the commercial vehicle for the launch, orbital operations, and landing phases of the flight and a fully trained NASA pilot would conduct the approach to the ISS, docking, and undocking operations; in this case, NASA astronauts would still need to be trained in emergency descent and rendezvous operations.)

One limitation of the taxi model is that it is more efficient if the person who is flying to the ISS is fully trained in ISS operations and other requirements (such as Soyuz flight engineer training). Astronauts who are trained only or primarily to operate the crew transport spacecraft may not be able to contribute to the ISS mission but will still use important ISS resources. Another limitation may be that the taxi pilot will take away valuable mass and volume for resupply of and return from the ISS. If the spacecraft spends only a short time at the ISS, dropping off astronauts and picking up those who will return to Earth, the spacecraft operator will not use ISS resources, but the spacecraft cannot be used as a "lifeboat" at the station.

Rental Car Model

In the "rental car" model, the provider would conduct launch and recovery operations but NASA would operate the spacecraft in flight with NASA astronauts. NASA's crew would thus require training in all aspects of flight operations. Because only the ISS-bound crew would journey to the ISS, this model reduces the logistical and habitation demands on the ISS. The spacecraft would remain at the station for emergency rescue and safe haven capability. The rental car model would require NASA's astronauts to receive additional ascent and rendezvous training (in contrast with the taxi model) in addition to providing technical, safety, and operational advice to commercial crew transport developers. This is similar to the Soyuz model.

Commercial Astronaut Training

The Astronaut Office anticipates that its personnel will aid commercial firms in developing training standards and training curricula for crews of commercial crew transports. NASA may furnish such support as part of its commercial launch services agreements or in cooperation with other government agencies. In addition, the FAA Office of Commercial Space Transportation will be responsible for licensing commercial astronauts, and NASA will have a role in informing requirements for selection and training.

Several commercial firms and FAA have hired former NASA astronauts as managers and consultants; these experienced fliers may reduce the need for members of the Astronaut Corps to participate extensively in development of commercial training curricula, FAA regulations, and licensing standards. However, there is also the possibility that the existence of the new firms could affect the NASA Astronaut Corps attrition rate as qualified personnel are attracted to private industry, which is seeking their skills.

Commercial Vehicle Development

During the Space Shuttle program, astronauts routinely followed the payload development, modifications to the orbiter fleet, and any technical efforts that affected shuttle system safety. Similar technical liaison between the Astronaut Office and commercial vehicle developers can be expected as NASA plans for crew transport operations on the private vehicles (see Box 2.1). NASA's astronauts may be tasked to act as liaison between companies that are building crew transports, providing operations and safety advice and keeping the agency abreast of technical progress. To preserve proprietary vendor information, NASA may need to assign a person to each of the commercial developers on a part-time basis; this would still constitute a substantial manpower commitment. The committee believes that there is a fundamental difference between having this role performed by a former NASA astronaut who is not actually to fly the spacecraft and by a current member of the Astronaut Corps who will be placing his or her life at risk in the vehicle. A current member of the Astronaut Corps, preferably with some spaceflight experience, will provide greater insight and credibility for a development vehicle that he or she will rely on in flight.

The near-term challenge for NASA is its role in ensuring the appropriate balance of "insight vs. oversight" for emerging commercial vehicles. This will of necessity be different from its role in connection with Soyuz or with a NASA vehicle development. The roles, rules, and processes are in development, including how the new commercial industry connects to the FAA role and the relationship between NASA, the FAA, and commercial industry.

As commercial crew transportation becomes available, NASA has several options for its use. They range from a commercial "taxi service" with NASA crew members ferried to and from the ISS to NASA's leasing or purchasing vehicles for NASA crew members to operate. In all cases, crew safety is of highest priority, and emergency crew return from the ISS requires NASA crew knowledge of and experience with all aspects of any commercial vehicle. Some aspects of potential NASA crew collaboration with commercial developers presented to the committee are the following:

- Perform unplanned activities and resolve emergency situations during all phases of flight.
- Provide input and recommendations on commercial crew spacecraft and operations.
- Collaborate in development of crew interfaces.
- Participate in testing, evaluation, and fit-checks.
- Participate in development of selection and training plans.
- Assist with procedures and operations development and verification.
- Raise concerns regarding safety and adherence to requirements (such as human ratings requirements).
- Participate in NASA programmatic activities (such as boards and panels) that require crew input and expertise.
- Rendezvous, proximity operations, and docking training.

NASA DEVELOPMENT PROGRAMS

NASA is developing a heavy lift booster and beyond Earth orbit-capable spacecraft. Termed the Space Launch System (SLS) and Multi-purpose Crew Vehicle (MPCV), respectively, these two systems are designed to enable the United States to conduct missions to the vicinity of the Moon, Lagrange points, near-Earth asteroids, and eventually the Mars planetary system.

As those efforts grow in scope and maturity, the Astronaut Corps will play an increasingly active role in their development and maturation. By 2015, at least a half-dozen members of the Astronaut Corps will be needed to follow and advise at the field center and program level.

During initial testing, the SLS and MPCV may be used to transport crew and cargo to and from the ISS, moving on to system demonstration first in high Earth orbit and then operationally throughout the Earth-Moon system. Lunar orbit sorties may be followed by Lagrange point expeditions, beyond Earth orbit assembly and servicing tasks, and eventual expeditions to near-Earth asteroids. Mission durations will range from a few weeks initially to multimonth asteroid expeditions.

To prepare for such ambitious operations (the first in nearly 50 years of beyond Earth orbit), active astronauts

BOX 2.1 Commercial Crew and Cargo Programs and Development—NASA and Non-NASA

To develop the next set of launch vehicles, capsules, and launch services after the end of the Space Shuttle program, NASA's Commercial Crew and Cargo Program Office (C3PO) has set up multiple contracts with aerospace companies based on design and performance milestones. NASA is using $500 million in Space Act Agreements for commercial cargo transportation and $50 million for the first round of commercial crew development agreements to fund these companies. NASA first began funding companies to provide resupply services to the International Space Station under its Commercial Orbital Transportation Services (COTS) and committed $269.3 million for its second round of Commercial Crew Development (CCDev) contracts awarded on April 18, 2010.

The following companies have received COTS and/or CCDev Space Act Agreements from NASA to develop spacecraft. This section focuses only on major hardware developments, not on Space Act Agreements for components of launch vehicle or cargo/crew capsule systems.

- *Blue Origin* is developing its New Shepard vehicle, a vertical take-off/landing vehicle designed to take humans on sub-orbital spaceflights, under both the CCDev 1 and 2 awards. On May 13, 2006, Blue Origin successfully test fired the Goddard vehicle, which reached a maximum altitude of 235 feet and landed safely within 50 seconds. Goddard is the precursor vehicle to the New Shepard. The New Shepard will be able to house both experiment racks for microgravity or other types of research and astronauts simultaneously. According to the company, "three or more positions" can be used by experiment racks or astronauts. Blue Origin expects to have its first opportunities for experiments that require the accompaniment of a researcher astronaut to be available in 2012.
- *Boeing* received CCDev 1 and 2 awards from NASA to develop its Crew Space Transportation (CST)-100 capsule, which will be able to carry seven astronauts to the ISS and the Bigelow Aerospace Orbital Space Complex. The CST-100 will be able to launch atop the Atlas, Delta, and Falcon 9 rockets and potentially other launch vehicles. The capsule will be able to operate autonomously for 48 hours while on orbit and take 6 hours to land (nominally on land, but water-landing contingency is possible) after undocking of the orbital platform. Boeing has already developed the pressurized structure and a crew module mockup as of February 2011. Boeing emphasizes the use of "heritage hardware" in the CST-100's design, including an Apollo-heritage parachute system, an abort system that uses existing components, and an airbag landing system from the Crew Exploration Vehicle/Orion. (See Figure 2.1.1.)

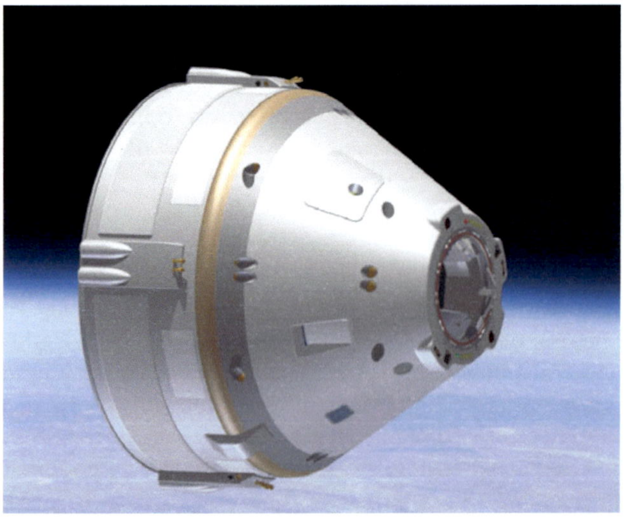

FIGURE 2.1.1 The Boeing CST-100 capsule.
SOURCE: Courtesy of the Boeing Company.

• *Orbital Sciences Corporation (Orbital)* is developing the Taurus II launch vehicle and Cygnus capsule for transporting cargo from Earth to the ISS, as well as disposing of space station waste. Orbital is contracted to deliver 20,000 kg of cargo to the ISS over the course of eight missions between 2011 and 2015. The Cygnus capsule is a pressurized cargo module, similar to the Multi-Purpose Logistics Modules already flown to the ISS. However, Cygnus is not designed to survive reentry of Earth's atmosphere, and will not be used to transport material from the ISS back to Earth. (See Figure 2.1.2.)

FIGURE 2.1.2 The Cygnus Advanced Maneuvering Spacecraft pictured with the International Space Station. SOURCE: Courtesy of Orbital Sciences Corporation.

• *Sierra Nevada Corporation* is developing the Dream Chaser, a fully reusable spacecraft based on the design of NASA's HL-20 lifting body vehicle, under CCDev 1 and 2 awards. The Dream Chaser is designed to launch atop the Atlas V launch vehicle and will return to Earth via conventional runway landing, which will allow it to bring materials back to Earth from the ISS. It will be capable of carrying seven astronauts and cargo to the ISS. The Dream Chaser is being designed so that each craft can be used for 50 to 100 missions. The company plans on placing a Dream Chaser into orbit by 2014 (Figure 2.1.3).

FIGURE 2.1.3 The Sierra Nevada Corporation's Dream Chaser lifting body shown traveling in space. SOURCE: Courtesy of Sierra Nevada Corporation.

• *Space Exploration Technologies (SpaceX)* is developing its Falcon 9 launch vehicle and Dragon capsule, which it will use to transport cargo to the ISS under its COTS agreement with NASA. Like Orbital Sciences Corporation, SpaceX is contracted to deliver 20,000 kg of cargo to the ISS over at least 12 missions with the option to order additional flights. The Dragon capsule is designed to survive re-entry into Earth's atmosphere and will be used for transporting materials back to Earth. On December 8, 2010, SpaceX successfully launched

BOX 2.1 Continued

a Falcon 9 and Dragon capsule. The Dragon capsule orbited Earth, survived re-entry, and was retrieved without incident. Dragon is designed to carry up to 6,000 kg to low Earth orbit (LEO) and 3,000 kg from the ISS back to Earth. The Dragon is for dual cargo/crew use and can accommodate seven passengers in its crew configuration. SpaceX received both CCDev 1 and 2 awards to launch astronauts to the ISS. In addition to the Falcon 9, which can launch up to almost 10,000 kg to LEO, SpaceX announced in April 2011 that it is beginning development of its Falcon Heavy launch vehicle, which will be able to launch payloads weighing over 53 metric tons to LEO, including interplanetary spacecraft. SpaceX cargo resupply missions to the ISS are slated to begin in 2012 (Figure 2.1.4).

FIGURE 2.1.4 The Dragon spacecraft in orbit. SOURCE: Courtesy of Space Exploration Technologies.

The following companies do not have Space Act Agreements with NASA but are developing hardware on a scale comparable with that previously listed under NASA COTS or CCDev development:

- *Bigelow Aerospace (Bigelow)* is developing an orbital space station complex that will combine its BA 330 and Sundancer inflatable modules into a single orbiting platform. The BA 330 inflatable space station module,

will assume a major role in development and flight testing, which will lead to a substantial technical and training commitment in addition to ISS crew expeditions. Initial astronaut advice on operations and cockpit and cabin design will be followed by integral involvement by members of the Astronaut Corps in ground system testing and then a series of crewed test flights.

By 2020 or 2025, according to present NASA planning, a sizable portion of the Astronaut Corps will be involved in preparations for beyond-Earth-orbit expeditions, including extensive field exploration on the surfaces of asteroids or the lunar surface. The possible missions include:

- Lunar orbit demonstration and survey.
- Geosynchronous satellite servicing.
- Hardware assembly at Earth-Moon Lagrange points.
- Servicing and assembly missions to Sun-Earth Lagrange point L2.
- Near-Earth asteroid expeditions.

Experience gained from those missions would enable eventual exploration of the martian moons and eventual expeditions to the surface of Mars. Astronauts will be required to demonstrate skills in ascent, entry, and rendez-

which will be able to house up to six astronauts in its 330 m^3 of space on an undefined long-term basis, will be serviceable by the Boeing CST-100. Before the launch of the BA 330, Bigelow will launch the smaller Sundancer inflatable module into LEO in 2014. Sundancer will be able to accommodate up to three astronauts on a long-term basis, and as many as six for short-term visits. Sundancer will have an internal volume of 180 m^3.

• *Lockheed Martin* is developing the Orion Crew Exploration Vehicle, which will serve as the reference design for NASA's multi-purpose crew vehicle (MPCV), which NASA is working on at Johnson Space Center. In particular, NASA is ensuring that the MPCV requirements are of the same scope as the ones that Lockheed Martin is using on Orion. Orion will be able to carry four astronauts to the Moon or six astronauts to the ISS and can sustain a crew mission independent of the ISS for 21.1 days. Lockheed Martin is also developing an Orion spacecraft and space operations simulation center at its Denver facilities. According to Lockheed Martin, the Orion vehicle is on schedule to conduct its first orbital flight test in 2013 and to provide initial operational flights by 2016 (Figure 2.1.5).

FIGURE 2.1.5 Artist's rendering of the Multi-Purpose Crew Vehicle in space. SOURCE: Courtesy of NASA.

vous piloting; docking operations; proximity operations around large space structures or asteroids; and descent to lunar and asteroid surfaces, as well as Mars.

The FCOD's Strategic Plan[7] responds to possible future needs by citing the following organizational objectives:

• Maintain adequate numbers of crew members with appropriate skills available to support all U.S. human spaceflight activities.
• Support advanced human spaceflight program development.
• Provide crew members for staff-critical program activities.

Given the dynamic environment, NASA crew task and skill requirements will not decrease in the foreseeable future. That is especially valid as NASA crews need to be prepared to conduct spaceflight operations during all phases, including unplanned and emergency situations. In particular, to fulfill its role of protecting the ISS as a valuable asset and ensuring safe operations, NASA will retain the responsibility for setting standards for procedures, training, and equipment for all docking and undocking operations.

[7] NASA, *Johnson Space Center Flight Crew Operations Strategic Plan*, September 2006, NASA Johnson Space Center, Houston, Tex., p. 12.

SUMMARY

As it examined the current status of the Flight Crew Operations Directorate and its Astronaut Office, the committee was impressed by the challenges of providing complex training for an Astronaut Corps that must both support long missions to the ISS and travel extensively to the facilities operated by the international partners. The committee was also impressed by the quality and professionalism of the Astronaut Corps and its management by the Astronaut Office. The committee noted that the Astronaut Corps is already experiencing the strains of downsizing and may have reached the minimum number of astronauts required to support current ISS commitments. The uncertainty imposed by the transition from shuttle to the ISS, such as the inability to predict attrition rates, makes it more difficult to predict how many members of the Astronaut Corps NASA will require. The committee concluded that the best way to navigate the transition was for the Astronaut Office to maintain its existing Astronaut Corps staffing model but to increase its margin from the current 25 percent.

The committee concluded that the Flight Crew Operations Directorate and its Astronaut Office currently support six tasks and should continue to do so. The most important is providing well-trained members of the Astronaut Corps to support the NASA flight manifest. The NASA Astronaut Corps is a resource, developed and refined over the last half-century, that is crucial for current U.S. space capabilities and for the commercial, national, and international future of the United States in space. Consequently, it is even more important in time of transition and uncertainty that the talent level, diversity, and capabilities of the Astronaut Office be sustained at the current task and skill levels. The limitations imposed by flying Russian spacecraft—including height restrictions and language skills—affect the diversity of the Astronaut Corps. In addition, the committee noted that astronauts who are no longer maintained in flight status still possess valuable experience and skills and are assigned other leadership roles, often outside the Astronaut Corps.

NASA has spent many decades in developing a proven human spaceflight capability, and changes in this capability should be made with great care and with a focus on human safety and mission success. In addition to current training, computer-based models and simulators that commercial space companies develop and more aggressive approaches to ground training may be incorporated in the future. As NASA moves to international cooperation as the standard for human spaceflight, and as it integrates commercial capabilities now in development into its day-to-day operations, the proven approaches to crew training currently in use by NASA serve as a firm foundation on which to prepare for an otherwise uncertain future (Figure 2.22).

FINDINGS AND RECOMMENDATIONS

Finding 2.1a. NASA's current Astronaut Office's role is to support six tasks (in priority order):[8]

1. Provide well-trained spaceflight operators to support the NASA flight manifest.
2. Provide ground support personnel for tasks required specifically to support the NASA flight manifest.
3. Provide support for new program development, ranging from development of relatively small payloads and equipment to development of whole new spacecraft designs.
4. Provide operational knowledge and corporate memory of human spaceflight.
5. Provide for collaboration with other government and private organizations as needed and directed by NASA.
6. Provide support for public and educational outreach to society.

The first task is the one in FCOD's model that drives the size of the Astronaut Corps—the number of astronauts qualified to fly in space. But the demands of tasks 2 through 6 add to the workload. The committee supports these roles as a proper use of an important core capability both now and into the future.

Management (inactive) astronauts serving in civil service positions in the Astronaut Office provide supplemental support for tasks 2 through 6. They do not use training assets except as instructors, evaluators, mentors, or providers of expertise; are ineligible for flight; and do not provide a reserve capacity for flight assignments.

[8] NASA identified tasks 1, 2, 3, 5, and 6; the committee has added task 4.

FIGURE 2.22 A trail of smoke is seen at dawn as STS-131 launches from Kennedy Space Center in Florida on April 5, 2010. SOURCE: Courtesy of NASA.

Finding 2.1b. Although NASA's human spaceflight program and its post-shuttle crew requirements have not been well defined beyond operation of the ISS, the sizing of the Astronaut Corps to meet ISS crew requirements has been well modeled by using ISS crew selection, training and flight recovery times, and a plan for post-shuttle force reduction.

Finding 2.1c. Astronaut anthropometric (physical size) limitations for flying in the Soyuz limit flexibility in crew assignments in response to contingencies.

Conclusion 2.1. On the basis of its assessment of known and potential needs, the committee concluded that the currently projected minimum staffing target size for the active Astronaut Corps poses a risk to the U.S. investment in human spaceflight capabilities. The committee concluded that given the array of potential crew assignment constraints and uncertainty in future requirements, the Astronaut Corps appears to be sized below the minimum required. The committee notes that the current plan for the size of the Astronaut Corps does not have the flexibility to accommodate commercial, exploration, and new mission development tasks or unexpected increases in attrition.

Recommendation 2.1.

- The committee recommends that the factor for uncertainty used by the Astronaut Office in its model to determine minimum staffing requirements for the Astronaut Corps be increased above the current 25 percent, which is inadequate to provide sufficient flexibility to meet the current flight manifest requirements reliably.
- In addition to task 1, the Astronaut Office should maintain the staff required to accomplish tasks 2 through 6 as listed in Finding 2.1a.

Finding 2.2. In addition to the need to meet NASA requirements, there is an expectation on the part of commercial crew providers and the Federal Aviation Administration that FCOD expertise and capabilities will be available in the future.

Recommendation 2.2. NASA's Flight Crew Operations Directorate should continue to serve as a national resource for U.S. human spaceflight experience and knowledge. This resource should be

- Maintained to ensure appropriate staffing and training of the Astronaut Corps in support of the International Space Station manifest;
- Applied to the future development of NASA human spaceflight and exploration activities;
- Available to the emerging commercial space industry and the FAA; and
- Applied to support authorized agreements with international partners.

3

Post-Shuttle Spaceflight Crew Training Resources: Ground-Based Facilities and T-38N Talon Aircraft

Facing the end of the Space Shuttle program—not unlike the preceding culmination of the Mercury, Gemini, and Apollo programs—NASA and the Johnson Space Center Flight Crew Operations Directorate are confronting the challenge of transitioning training resources to focus on meeting the needs of the International Space Station and future commercial space systems. In the transition process, it is imperative that NASA retain resources to prepare astronauts to continue to achieve safety and mission success through the ISS and future space systems. To address the future resources needed, the committee was tasked to examine the requirements for crew-related ground-based facilities in the post-shuttle era. The committee was also asked to determine whether the Astronaut Corps's fleet of training aircraft is a cost-effective means of preparing astronauts for the requirements of NASA's human spaceflight program. To address those issues, the committee explored a variety of topics, including astronaut training and proficiency requirements, operator skills and high-performance aviation training, current and post-shuttle resources, and evolving training methods.

TRANSITION TO A POST-SHUTTLE ASTRONAUT CORPS

The post-shuttle Astronaut Corps faces several possible futures. One is an ISS-only scenario, in which astronauts will serve tours aboard the ISS through 2020 or even through 2025 and serve as crewmembers on transport vehicles to and from the ISS. Managers must also anticipate that commercial spacecraft developers will require NASA assistance and potential personnel "loans" to achieve rapid NASA and Federal Aviation Administration (FAA) certification for transportation to and from the ISS. To promote safety and mission success and to avoid duplication and parallel training establishments among financially constrained commercial firms, it may be appropriate to consider the Astronaut Office and Astronaut Corps as a national asset that must be capable of supporting additional staffing requirements to assist such firms. Another such future could include the addition of a beyond-Earth-orbit flight program between 2016 and 2020 that would also require additional astronauts to aid in development and flight testing of the new system. Furthermore, eventual expeditions to deep space, the Moon, or nearby asteroids beyond 2020 could also be staffed by a slightly larger corps.

The training of astronauts in the post-shuttle era points toward a shift to skills-based training, as on the ISS, rather than the task-based regimen of the shuttle era. In the past, shuttle missions 1-2 weeks long lent themselves to intensive training focused on a well-defined set of tasks in the specific mission flight plan. On ISS expeditions lasting months, unanticipated pre-flight and in-flight changes in planned activities have often occurred, so members

FIGURE 3.1 The International Space Station as seen by space shuttle *Discovery* on March 25, 2009, after undocking. SOURCE: Courtesy of NASA.

of the Astronaut Corps must possess a flexible, broad base of skills, including EVA, robotics, payload operations, in-flight maintenance, and potential emergency responses. The ISS is a large, complex orbiting laboratory facility that has many systems and tasks (Figure 3.1).

ISS training also features a sizable international component that deals with spacecraft, modules, and hardware from Russia, Europe, Japan, and Canada. The committee assumes that the international aspect will continue during the era of commercial spaceflight and remain a major facet of beyond-Earth-orbit operations if they occur.

TRAINING AND PROFICIENCY REQUIREMENTS

A distinctive aspect of participating in spaceflight is that each astronaut is expected to arrive in space having already developed the individual skills necessary for real-time decision making in an operational environment. Exercising such skills relies on innate capability to set priorities among a variety of factors, from a keen appreciation of the risks associated with decisions that affect personnel safety to a sound understanding of the design specifications that affect vehicle reliability to the trade-offs that affect operational efficiency. Three prototypical decision-making behaviors provide a reasonable benchmark: skills-based, rules-based, and knowledge-based.[1] That benchmark is used in other safety-critical, time-critical domains, such as commercial aviation and the civilian nuclear industry[2] (see Appendix A). Likewise and arguably even more prominently, the astronaut training program must prepare astronauts to apply those different behaviors at appropriate times. Training astronauts to a suitable level of proficiency in the decision-making behaviors requires various training methods and facilities.

[1] J. Rasmussen, Skills, rules, knowledge; signals, signs, and symbols, and other distinctions in human performance models, *IEEE Transactions on Systems, Man, and Cybernetics* 13:257-266, 1983.

[2] Institute of Nuclear Power Operations, *Human Performance Reference Manual,* Atlanta, Ga., October 2006.

CREW TRAINING GROUND FACILITIES AND THE SPACE SHUTTLE'S RETIREMENT

The committee's second task was to address the agency's ground training facilities. As NASA, the FCOD, and the Astronaut Office enter the post-shuttle era, astronaut training will be exclusively devoted to International Space Station missions. Throughout the era of the shuttle, NASA has possessed a vast ground infrastructure to support shuttle processing and on-orbit operations. In particular, the Johnson Space Center possessed many shuttle-related training facilities. They included the large mock-ups that bear a resemblance to the orbiters and are visible to public tours of the space center but also numerous other facilities and pieces of equipment, such as the Shuttle Mission Training Facility's Guidance and Navigation Simulator. As this report was being written, NASA was in the process of decommissioning many of its simulators and trainers; most will be donated to museums or universities. See Table 3.1 for the disposition of the various NASA-operated shuttle ground facilities and trainers.

In addition, as this report was being written, the committee was informed that NASA had been contacted by potential commercial crew providers about leasing portions of some NASA facilities, such as the large Building 9 facility at Johnson Space Center, for commercial crew training, and having NASA astronauts aid in curriculum and simulator development. The discussions were still in a preliminary stage, and the committee had little information about them.

TABLE 3.1 Disposition of NASA's Shuttle Ground Training Facilities

Facility	Status Post-Shuttle	Final Location
Shuttle Mission Training Facility (SMTF) fixed base	Decommission	Adler Planetarium
SMTF motion base	Decommission	Texas A&M University
SMTF Guidance and Navigation Simulator	Decommission	Wings of Dreams Aviation Museum
Single System Trainers (3)	Decommission	1 to Texas A&M University 2 as static displays at NASA Johnson Space Center (JSC)
Dynamic System Trainers	Remain operational; support International Space Station (ISS) robotics and VV rendezvous training	
Payload Trainer	Decommission	NASA JSC
Network Simulation System	Decommission	NASA JSC
Shuttle Engineering Simulator Dome	Decommission (committee recommends evaluating need)	NASA JSC
Space Station Training Facility	Remains operational; supports ISS training	
Space Station Mockup Training Facility	Remains operational; supports ISS training	
Part Task Trainer	Remains operational; supports ISS training	
Full Fuselage Trainer	Decommission	Seattle Museum of Flight
Crew Compartment Trainer (2)	Decommission	1 to Air Force Museum 1 to Smithsonian Institution
Neutral Buoyancy Laboratory	Remains operational; supports ISS training	

Systems Engineering Simulator Dome

One facility currently used for shuttle and ISS robotics and rendezvous training, the Shuttle Engineering Simulator (SES) Dome, is slated for decommissioning. The SES Dome is used by NASA to provide crew training and engineering analysis of on-orbit operations. It contains an orbiter aft cockpit mockup in the dome; an ISS cupola with a robotic work station; an orbiter forward cockpit integrated with ascent and entry simulation and launch and landing site scenes; an Orion crew station mockup; a reconfigurable operational cockpit; and the stand-alone Dynamic Skills Trainer for robotics, rendezvous operations, and ascent and entry training.[3] The SES Dome is able to provide several areas of training, such as virtual reality laboratory scenes, plume modeling, high-fidelity berthing and docking contact and mechanism models, and vehicle guidance, navigation, and control.

Robotics training and commercial rendezvous and docking procedure development will still be required in the near term. Resupply of the ISS is vital to its continued operation and will involve not only the Progress, ATV, and HTV spacecraft already developed but commercial vehicles, such as SpaceX's Dragon and Orbital's Cygnus-1, which are in development, and possibly additional vehicles as well. Because the SES Dome is a unique facility that is not replicated elsewhere within NASA, the committee believes that it may be a valuable asset that NASA should evaluate for future use.

In addition to the shuttle facilities and trainers, NASA has numerous assets that are required for training astronauts in use of the International Space Station. Current NASA plans require that the facilities will be retained; for example, the Neutral Buoyancy Laboratory is still required for training astronauts for spacewalks.

Neutral Buoyancy Laboratory

The Neutral Buoyancy Laboratory (NBL), adjacent to Ellington Field at NASA's Johnson Space Center in Houston, Texas, trains astronauts for extravehicular activity. The facility puts astronauts into a pool of water 40 feet deep, 102 feet wide, and 202 feet long that holds 6 million gallons. With proper weighting of their spacesuits, astronauts experience neutral buoyancy, enabling them to move their suits, tools, and equipment in a manner close to the orbital free-fall environment. Although gravity still makes dropped tools fall to the bottom of the tank and astronauts still know which way is "up," they are able to move about and work on full-scale mockups of the shuttle and space station. In preparation for a 6-hour EVA, an astronaut might spend as much as 10 times that in underwater rehearsal, acquiring the skills and discipline necessary to accomplish EVA objectives.

Shuttle astronauts scheduled for multiple EVAs on an ISS assembly mission typically spent more than 200 hours training in the NBL. ISS expedition crews train in the NBL to develop EVA skills proficiency and to learn in detail the layout of the ISS structures that they will encounter while maintaining the outpost in orbit. While working for 6 hours in the depths of the NBL, astronauts are immersed in the physical and mental environment of a free-fall EVA (Figures 3.2, 3.3, 3.4, and 3.5).

The committee queried NASA regarding studies of high-fidelity simulators to meet the Missions Operations Directorate or Flight Crew Operations Directorate crew training requirements. NASA informed the committee that three main factors drive the selection of simulator fidelity: the criticality and complexity of the task to be performed, the effectiveness of the different fidelity options, and budget. NASA also gave the committee the results of the Constellation Training Facility Trade Study, particularly the Orion Part Task Trainer, as an example of recent work that the agency has conducted. With the possible exception of the SES Dome, which may need to be retained for continued ISS training, the internal NASA facilities appear well suited to preparing ISS astronauts for safe and successful missions.

In addition to NASA facilities in the United States, ISS training facilities are distributed around the world: at Russia's Gagarin Cosmonaut Training Center near Moscow, Russia; at the European Space Agency's Astronaut Center in Cologne, Germany; at the Japanese Aerospace Exploration Agency's Tsukuba Space Center; and at the Canadian Space Agency. The international distribution of facilities not only complicates NASA astronaut training but introduces inefficiencies. For example, a NASA astronaut qualifying as a Soyuz flight engineer will spend 49

[3] Available at http://www.nasa.gov/centers/johnson/engineering/flight_design/systems_engineering_simulator/index.html.

FIGURE 3.2 Astronaut training in the Neutral Buoyancy Laboratory. SOURCE: Courtesy of NASA.

FIGURE 3.3 STS-63 astronauts Bernard A. Harris and C. Michael Foale prepare to exit *Discovery's* airlock for a spacewalk. SOURCE: Courtesy of NASA; GPN-2006-000022, available at http://grin.hq.nasa.gov/.

FIGURE 3.4 EVA mission specialist Pierre Thuot is seen making an attempt to capture the Intelsat VI communications satellite with the satellite capture bar on the Remote Manipulator System. SOURCE: Courtesy of NASA; GPN-2000-001096, available at http://grin.hq.nasa.gov/.

weeks in the United States, 2 weeks in Europe, 31 weeks in Russia, 7 weeks in Japan, and 2 weeks in Canada. In addition, 12 percent of the time will be spent purely on travel from one training location to another[4] (Figures 3.6 and 3.7).

HIGH-PERFORMANCE AIRCRAFT AND ASTRONAUT TRAINING

The committee's third task was to address the agency's fleet of astronaut training aircraft. The Space Shuttle program required several aircraft specifically for training astronauts on how to land the orbiter. Those aircraft are being retired or directed for other uses and were not addressed by the committee. The task refers instead to the fleet of T-38N Talon two-seat training aircraft (Figure 3.8).[5]

---

[4] NASA, *Certification of Flight Readiness Process Document: International Space Station Program,* SSP 50108, Revision C, NASA Johnson Space Center, Houston, Tex., November 2006; NASA International Training Control Board, *International Space Station Multilateral Advanced/Increment-specific Training Plan (MA/ITP),* Version 7.0 Baseline, SSP 50170, NASA Johnson Space Center, Houston, Tex., March 2010; NASA Flight Crew Operations Directorate, *Flight Crew Operations Space Flight Preparation Plan,* CA-QMS-001, Revision F, NASA Johnson Space Center, Houston, Tex., August 2010; P.A. Whitson, NASA Astronaut Office, "Presentation to the NRC Committee on Human Spaceflight Crew Operations," presentation to the Committee on Human Spaceflight Crew Operations, January 6, National Research Council, Washington, D.C., 2011, pp. 55-57 and 64; P.S. Hill, NASA Mission Operations Directorate, "MOD Crew Training," presentation to the Committee on Human Spaceflight Crew Operations, March 1, National Research Council, Washington, D.C., 2011.

[5] The T-38N Talon aircraft is the flight readiness training aircraft specifically used for NASA's spaceflight training needs. The Air Force uses the T-38C model, which is slightly different. The T-38N aircraft includes differences in communications and navigation, such as the addition of weather radar, a data link weather system, the terrain avoidance and warning system, the terminal collision avoidance system, GPS with localizer performance and vertical guidance approach capability, and redesigned electrical, inlet, ejector nozzle, and flight management systems. There are 7 Block 2 aircraft of the T-38N and 14 Block 3 aircraft. The Block 2 version from 1990 included the first "glass cockpit"

FIGURE 3.5 Astronauts G. David Low and Peter J.K. Wisoff attached to the end of the space shuttle's robotic arm. SOURCE: Courtesy of NASA; GPN-2000-001073, available at http://grin.hq.nasa.gov/.

In addressing this task, the committee considered several questions: What is the role of the T-38N aircraft? Why is such an aircraft necessary? Does its use by astronauts reduce the risk to the nation's space effort? Could alternative aircraft fill the same role more cost effectively? Could a simulator or some combination of aircraft and simulators perform the same role more cost effectively?

The committee determined that the T-38N aircraft fulfill several roles for the Astronaut Office. One role is to enable the Astronaut Office to recruit and maintain military test pilots for the Astronaut Corps. Military test pilots who are selected by the Department of Defense (DOD) and NASA to become astronauts are expected to maintain

design and safety upgrades. The Block 3 version from 2007 incorporated an electronic flight instrumentation system and several additional safety upgrades in compliance with NASA and FAA requirements. The T-38N has recorded no Class A mishaps (involving fatality, total disability, or more than $1 million in damage) in 100,000 flight hours since 2000.

FIGURE 3.6 Astronaut Ellen Ocha simulates an emergency egress procedure at Johnson Space Center's Mockup and Integration Laboratory. SOURCE: Courtesy of NASA; GPN-2000-001068, available at http://grin.hq.nasa.gov/.

FIGURE 3.7 The International Space Station includes equipment and personnel from many countries, and this requires NASA astronauts to travel to distant locations to train on equipment and requires international partner astronauts to train in the United States. Here a Russian Federal Space Agency cosmonaut participates in an extravehicular activity as a part of Expedition 17 on July 15, 2008. SOURCE: Courtesy of NASA.

a minimum number of high-performance flight hours each month as provided by NASA through its flight operations at Ellington Field. If the high-performance proficiency jet training is eliminated, there is the possibility that the DOD will no longer provide military astronauts with flight test expertise, or the pool of exceptionally qualified pilots may decrease when they recognize that they may not have the opportunity to maintain flight proficiency; this could affect their military flight currency requirements, later promotions, and reintegration into their parent service.

Military test pilots will remain an important component of the Astronaut Corps, and the Astronaut Office has indicated that continuing to recruit them has high priority after the retirement of the space shuttle so that it can take advantage of their expertise in the operation of complex equipment in a high-stress and dangerous environment. However, even if that were not the case, the T-38N aircraft serves an important role in providing spaceflight readiness training (SFRT).

NASA has stated that a purpose of spaceflight readiness training is to maintain "a pervasive, high-performance aviation safety culture that has become the cornerstone for and continues to build upon an analogous space operations safety culture." The committee agrees that such training is valuable, particularly for a diverse population of new astronauts.

New U.S. astronauts come from a wide variety of backgrounds and professional experience—from pilots, scientists, and engineers to medical doctors and educators. For example, the June 2011 Astronaut Corps of 61 included 50 men and 11 women; 17 are military, and 44 are civilian.

To give this diverse group of professionals a common experience base in a highly demanding operational environment, NASA exposes the Astronaut Corps to regular flights in high-performance aircraft (currently the Northrop T-38N). On most sorties, pilot astronauts or NASA instructors occupy the T-38N front seat, and a non-

FIGURE 3.8 T-38N jets in flight over NASA's Dryden Flight Research Center in California. SOURCE: NASA Dryden Flight Research Center. NASA/Jim Ross; ED07-0222-06, available at http://www.nasa.gov/centers/dryden/multimedia/.

test-pilot astronaut student flies from the rear cockpit. When necessary, pilot astronauts receive instruction in the front seat from an instructor seated in the rear. The instructor pilots provide ground-based systems, procedural, and emergency training, sometimes using NASA's T-38N cockpit simulator. In the air, the instructor pilots conduct flights to fulfill the SFRT training syllabus, addressing mission objectives of increasing complexity and difficulty for each astronaut or astronaut candidate. The instructor pilots evaluate astronaut in-flight performance and administer formal inflight evaluations to both pilot and non-pilot astronauts.

The requirements for spaceflight readiness training are outlined in the *Astronaut T-38 Space Flight Readiness Training Syllabus.*[6] In particular, the committee notes that the Mission Specialist Annual Qualification Check listed in the syllabus indicates the broad range of activities that non-pilots are required to train for with the admonition to "RECOGNIZE ANY UNSAFE SITUATION/CONDITION!!!!" (emphasis in original). (See Box 3.1.)

As a review of the syllabus demonstrates, mission specialists do not serve as "passengers" in the back seat of the T-38N; they are part of the aircrew and have such responsibilities as operating communications and navigation equipment at night and under adverse weather conditions. Once astronauts are capable high-performance aircraft crew members, spaceflight readiness training aids in further developing Crew Resource Management (CRM) skills. CRM skills are those that can assist a cockpit crew in assessing emergency situations, recognizing task saturation, and avoiding critical errors in collective judgment that might result in an accident. In addition to practicing CRM

[6] NASA Aircraft Operations Division, *Astronaut T-38 Space Flight Readiness Training Syllabus,* AOD 37515, Rev. C, NASA Johnson Space Center, Houston, Tex., July 2000.

in the shuttle mission simulators, each T-38N flight exposes the two crew members to a unique laboratory for practicing and developing skills in interpersonal communication, leadership, and decision making. Exposure to the unscripted flight environment challenges astronauts to ignore distractions, set priorities for decisions, and get the best talents from the team to conduct a safe flight.

The agency's goal is to develop essential spaceflight operator skills that will enable members of the Astronaut Corps to perform successfully in an unpredictable and dynamic environment where decisions have real-world survival consequences. T-38N flying has delivered a level of skill and experience that has proved to be a successful and acceptable standard of training for NASA astronauts in the shuttle, Soyuz, and ISS environments. Such preparation is directly applicable to spaceflight events, such as cargo vehicle "free-flyer" robotic capture, EVA operations, Soyuz ascent and re-entry, and ISS emergencies (which have included fire, depressurization, toxic contaminant release, and electrical failure). NASA's experience in ensuring safety and mission success has shown that ground-based knowledge training alone is not sufficient to produce high confidence that a crew member can perform under duress in orbit. Operations skills, rapid response, successful "triage" under the high pressure of unexpected malfunctions, and cognitive processing abilities must be trained and developed. As flight hours fall below the current syllabus levels, astronaut cockpit performance suffers; NASA's instructor pilots report degraded physical skills (hand-eye coordination) and decision-making skills. The instructors consider the current flight time levels and syllabus requirements as the minimum acceptable.

Detailing its resources available for knowledge, skills, cognitive, and rapid response training, NASA asserts that only a high-performance aircraft directly addresses the desired traits to provide high confidence in ensuring safety and mission success. As Table 3.2 indicates, the T-38N is currently the only Astronaut Corps training resource that combines the categories of skills, cognitive control, and rapid response training essential for an ISS emergency—for example, Soyuz ascent and entry, rendezvous and docking failures, complex ops (free flyer capture), and fire.

NASA is not the only spaceflight agency that recognizes the need for such training. For example, during the training for Spacelab 1 (STS-9) in 1983 and German Spacelab Mission D-1 (Deutschland-1) STS-65A in 1984-1985, all three German astronauts and the Dutch astronaut were required to attain pilot's licenses by the German Aerospace Agency, and it provided a plane for them to fly. German Aerospace Agency management felt strongly about this flight experience before shuttle flight because it created a real-time operational environment for persons who had spent most of their career in the laboratory. Later, U.S. partners on the ISS (the Japanese Space Agency, JAXA, the Canadian Space Agency, and the European Space Agency) contracted with NASA for T-38N SFRT. According to NASA, the Russian cosmonauts are provided aircraft flight time, and, because the Chinese select their taikonauts from active military pilots, they are presumably providing them with flight currency. The Russians provide their cosmonauts much less aircraft training time because of resource constraints (the transfer of the Gagarin Training Center from Russian air force to civilian control resulted in the loss of all their aircraft except a modified Tu-154), but they do provide some. In addition, the Russians provide every trainee cosmonaut, whether military pilot or civilian engineer, with supplemental stressful training in the form of parachute jumping.

Three other astronaut training resources provide "rapid response training" (for instance, the building 9 modules), but only the T-38N provides the urgency of life and death decisions. There is a fundamental difference between a training situation in which trainees can walk away from their mistakes and one in which they cannot.

Despite their impressive overall credentials, not all astronaut candidates come to NASA with operational skills and experience. High-performance aircraft training has instilled and expanded candidates' capability and competence in performing in a fast-paced spaceflight environment. The committee noted that commercial spacecraft developers also call for cockpit proficiency as the fundamental underpinning of training of new crew members.

The T-38N's safety record reflects both the quality of maintenance and the aircrew experience: no lives have been lost in nearly 40 years. Although two NASA pilots died in a January 1972 crash during an instrument approach in the fog, the last T-38 accident that caused the death of an astronaut occurred in 1967. The last T-38 major accident (Class A) occurred in November 1982, and involved no fatalities. Since then, NASA has flown more than 270,000 T-38N accident-free hours, compiling an impressive 0.00 mishap rate; this rate compares favorably with

BOX 3.1
Excerpt from *Astronaut T-38 Space Flight Readiness Training Syllabus*

MS ANNUAL QUALIFICATION CHECK
OBJECTIVE

The purpose of the MS annual qualification check is to ensure that the MS is proficient in assisting the pilot in normal and emergency situations. Crew coordination, checklist procedures, instrument procedures, and systems knowledge should be emphasized.

Expected Standards of Proficiency for Mission Specialist Crew Duty Day Extenders

NOTE *For details on expected aircrew proficiency e standards, see AOD 33869, T-38 Aircrew Proficiency Standards.*

- Be able to quote or write all boldface items from memory
- Have the following ops limits memorized: EGT flight limits, nozzle limits, minimum fuel, oil pressure limits, hydraulic pressure limits
- Calculate TOLD from the checklist
- Check weather, NOTAMs, and servicing availability if going cross-country.
- File a flight plan, copy clearances, get ATIS, and program flight plan in FMS
- Perform all preflight inspections including aircraft walk-around, parachute and ejection seat preflight
- Communicate with ATC (ideal), or understand radio transmissions (minimum)
- Navigate: select EHSI/EADI screens as requested by pilot or for the situation; have required navigation aids, displays, altitudes, headings, MDA's, and courses set in order to accomplish the SID/STAR/ en route navigation/approach
- Direct an inflight divert
- Verify correct aircraft configuration
- Compute final approach airspeeds
- Ensure completion of checklists (specifically, "Before takeoff," through "after landing")
- Have a basic knowledge of systems and emergencies
- Be able to locate emergency procedures in the PCL during flight, and execute them
- Have a thorough understanding of ejection system, lap belt, parachute, and oxygen system
- RECOGNIZE ANY UNSAFE SITUATION/CONDITION!!!!

PREFLIGHT BRIEFING

a. Complete Boldface and Ops Limits Test
b. Grade and review written exams
c. Brief FOD prevention
d. Brief Cabin Pressure Loss
e. Brief a procedure selected from the following list:
  - anti-ice
  - weather radar
  - approach categories
  - cold weather procedures
  - weather minimums

f. Brief an Emergency Procedure (EP) selected from the following list:
  - generator failure
  - Nav System failure
  - radio out
    - Hydraulic systems
    - Engine systems
    - landing gear systems

g. Standard Briefing from In Flight Guide
h. Ejection Seat Briefing (before or after the flight)

Prior to flight MS will:

- Check PIF
- Check Weather
- Check Notams
- File Flight Plan
- Calculate Take off and Landing Data

MANEUVER PROFILE - REQUIRED ITEMS

a. Manage comm and nav on instrument approaches.
1. Precision
2. Non-precision.

b Perform area maneuvers (if weather permits - optional)
1. Engine shutdown and relight
2. Stalls
3. Aerobatics (optional).

c. Simulated Emergency Procedures
1. Aircraft emergency
2. Divert

d. Perform visual patterns. ([international partners] demonstration - optional)
1. Heavyweight single-engine
2. Single-engine touch-and-go and go-around (climb to 2,000 feet AGL).
3. No flap
4. Minimum run.

DEBRIEF:

EJECTION SEAT BRIEFING

a. Ejection envelope (0 feet, 50 knots)

b. Pre-ejection
1. Oxygen
2. Parachute straps
3. Other items.

c. Ejection
1. Body position (head, elbows in, feet)
2. Trigger guard (trigger guard, leg guard movement).

d. Post-ejection
1. Beat the system
2. Oxygen
3. Check parachute canopy
4. Four line cut
5. Survival kit deployment
6. Landing, body position, parachute release.

e. Emergency ground egress

f. Parachute accessories
1. Survival kit (in parachute)
2. Beeper, radio
3. Strobe light.

g. Survival kit contents

SOURCE: NASA Aircraft Operations Division, *Astronaut T-38 Space Flight Readiness Training Syllabus,* AOD 37515, Rev. C, NASA Johnson Space Center, Houston, Tex., July 2000, pp. 97-99.

TABLE 3.2 Resources Capacity to Provide Astronaut Knowledge, Skills, Cognitive, and Rapid Response Training

Training Resource	Knowledge	Skills	Cognitive	Rapid Response
Building 9 Airlock	√	-	-	-
Building 9 Modules	√	-	-	√
Building 9 Racks	√	-	-	-
Building 9 SSRMS	√	√	-	-
Dynamic Skills Trainer	√	√	-	-
Neutral Buoyancy Laboratory	√	√	√	
Part-Task Trainer	√	√	-	-
Space Station Training Facility	√	√	-	√
Single System Trainers	√	-	-	-
Virtual Reality Laboratory	√	√	√	-
Medical	√	√	-	√
T-38N	-	√	√	√
NASA Extreme Environment Mission Operations	-	√	√	-
National Outdoor Leadership Seminar	-	√	√	-

NOTE: Many resources cover several categories, but the T-38N training aircraft is the only resource that NASA has that can provide skills, cognitive, and rapid response training. SOURCE: NASA Astronaut Office, "Ensuring the Readiness of the Astronaut Corps: A White Paper," NASA Johnson Space Center, Houston, Tex., March 25, 2011.

the U.S. Air Force rate of 1.09 (fiscal year [FY] 2005-FY 2009), the U.S. Navy rate of 1.28 (FY 2005-FY 2009), and the U.S. Marine Corps rate of 1.90 (FY 2005-FY 2009).[7]

It is particularly relevant to post-shuttle operations that ISS crew experience over the last decade has reflected how actual spaceflight anomaly responses correlate with emergency response preparation typically found in an aircraft environment. ISS astronauts and cosmonauts have had to react to more than 850 anomalies over 10 years of ISS operations that required critical, rapid responses and multitasking skills, for example,

- The Soyuz TMA ballistic re-entry after Expedition 6 on May 3, 2003, was caused by a flight control avionics failure. The ballistic re-entry mode subjected the crew to more than 8 Gs during re-entry and shifted the landing site more than 500 km up-range. The physical demands and strict, timely procedural discipline required during this potentially dire incident were similar to those demanded in T-38N training.
- During Expedition 16 (October 2007-April 19, 2008), a fray in a guide wire resulted in tearing between two photovoltaic panels during redeployment of a solar array. Ground teams worked rapidly with shuttle and station crews to develop repair procedures. Astronauts aboard the ISS fashioned structural reinforcements known as "cufflinks," and during a spacewalk they cut the damaged guide wire, freed the solar array, and installed the reinforcements to permit full extension of the array. The close ground-space coordination and critical safety procedures implemented to mitigate EVA hazards were similar to practices experienced in a high-performance aircraft environment.

[7] NASA Flight Crew Operations Directorate, "Presentation to the NRC Committee on Human Spaceflight Crew Operations," presentation to the Committee on Human Spaceflight Crew Operations, January 5, National Research Council, Washington, D.C., 2011; R.N. Clark, NASA Aircraft Operations Division "Presentation to the NRC Committee on Human Spaceflight Crew Operations," presentation to the Committee on Human Spaceflight Crew Operations, January 6, National Research Council, Washington, D.C., 2011.

- As Expedition 16 landed on April 19, 2008, the commander, a biochemist researcher, attributed her successful ability to execute emergency procedures as the Soyuz flight engineer during the return flight to Earth and to keep focused on tasks during a 7 G off-nominal ballistic re-entry to spaceflight crew readiness training in the T-38N. The commander was convinced that this unexpected high-stress situation might have ended less successfully if there had been no prior flight experience in private aviation or the SFRT program.
- During Expedition 22 (September 2009-March 2010), both primary and backup command and control computers failed aboard the ISS. Caution and warning alarms sounded, and control was lost over all core systems in the U.S. orbiting segment of the ISS, including communications. The crew responded with a complex recovery procedure that used a third computer and restored communications long enough to respond correctly when two successive failures occurred. The crew relied on training and systems knowledge to cut off an errant telemetry stream coming from the Columbus Laboratory. The required crew situational awareness, coordination between crew and ground, and concise communication during repeated outages were responses similar to those developed in high-performance aircraft training.
- On Expedition 23 in August 2010, an ISS external coolant pump failed during crew sleep, cutting ISS power by half and requiring rapid reconfiguration of essential core systems. The crew worked extensively with ground teams to obtain a stable ISS configuration and in the following weeks performed three critical EVAs to replace the pump and restore cooling. The required crew resource management, situational awareness, communications skills, and critical response under time pressure were similar to skills demanded in high-performance aircraft training.[8]

With such examples highlighted from the hundreds of anomalies that ISS astronauts encountered, the NASA approach to flight crew training is the result of 50 years of successful experience that has led to certified astronaut crews that conduct safe operations in a demanding, unforgiving, and hazardous environment. NASA's astronauts have included people who did not originate in a flight organization or career (such as Apollo scientist-astronauts, space shuttle mission specialists, ISS flight engineers, and educators), and such training brought them to a level of operational capability in which they could meet or exceed safety and mission success requirements.

INTERNATIONAL PARTNERS AND COMMERCIAL AVIATION TRAINING

Other countries besides the United States recognize the importance of some form of aviation training for their astronauts. Professional space crew members in Russia and China also draw from a pool of experienced pilots for the same reasons as the United States. Star City management is concerned that its 2009 transition from the Russian Air Force to the civilian Roscosmos has reduced its access to training aircraft. International Space Station partners Canada, the European Space Agency, and the Japanese Aerospace Exploration Agency (JAXA) provide pilot training for their crew members in their home countries, "reimburse" NASA to provide T-38N flight readiness training before mission selection, or both.

Flight training and the issuance of pilot licenses require demonstrated and evaluated proficient operation of aircraft in both nominal and stressful and abnormal situations: incipient stalls, loss of engine, loss of instrumentation, and so on. Although the systems of one aircraft may not be the same as those of another, the pilot's critical decision-making skills do transfer and constitute part of the evaluation process by a Federal Aviation Administration examiner. The Federal Aviation Administration has recently ruled that all commercial space vehicle crew members must have at least a valid FAA flight license if they are to be qualified to command commercial space vehicles.

Commercial pilots are able to re-certify in new planes via simulators. However, an Air Transport Pilot license requires 1,500 flight hours. Pilots are initially hired as "first officers" who are then positioned to gain hundreds or thousands of hours of flight experience under the supervision of a captain, and pilots remain subject to annual airplane check rides. George Nield, Associate Administrator for Commercial Space Transportation of the FAA,

---

[8] NASA Astronaut Office, "Examples of Actual Spaceflight Anomalies with Correlation to Training/Preparation for Emergency Response," attachment to "Responses to Questions for NASA on Space Flight Crew Issues," submitted to the Committee on Human Spaceflight Crew Operations, National Research Council, Washington, D.C., March 1, 2011.

informed the committee that he considered it "inconceivable" that private or commercial pilots would ever be licensed for operations without prior flight experience—simulator training is insufficient.

The reality of spaceflight operations is that crew members must demonstrate

1. Hand-eye coordination in fast-moving dynamic and stressful flight environments (high G, reduced pressures in pressure suits, claustrophobic spaces, and other physical and psychological stresses).
2. Critical decision making in the same environments (such as real-time nominal, off-nominal, and emergency operational procedures; problem solving; responding to unexpected events outside the training syllabus and procedures; scanning multiple computer displays; entering commands; and changing radio frequencies).
3. Ability to multitask in these environments and make triage decisions.
4. Ability to execute all of these tasks while communicating with other crew members and external communication sources (FAA in aviation and Mission Control in spaceflight) and demonstrating good Crew Resource Management (CRM) skills.

The committee was briefed by Bryan O'Connor, then the head of NASA's Office of Safety and Mission Assurance, on the role of crew training in NASA's overall safety and mission assurance process. O'Connor informed the committee that "flight crew operations is an integral part of formal agency risk management." He further stated that "Decisions involving crew safety risk require formal concurrence by the cognizant technical authorities, as well as formal consent to take the risk by the flight crew (and their supervisory chain) before the program, project, or operations manager may formally accept the risk." One input to the risk management process is the contribution of the SFRT program to crew training for rapid response to time-critical emergency situations under flight environment pressures.

The spaceflight readiness training requirement is not tied to a specific mission but is derived from safety and mission success requirements established by NASA Headquarters. Although that requirement is not expressly documented at the NASA Headquarters program level, it is developed by FCOD in response to the Headquarters-controlled safety and mission success requirements and embedded in the NASA JSC-level Certificate of Flight Readiness (CoFR) for safe operations of flight. This CoFR is required for any launch of a U.S. astronaut—whether on a U.S. spacecraft, such as the shuttle, or on a non-U.S. spacecraft, such as the Soyuz—and must be provided to NASA Headquarters before a launch is approved. The flight readiness review processes and timelines requiring the signed CoFR are identified in Figure 3.9 for the ISS and in Figure 3.10 for Soyuz.

The SFRT requirements are not mission unique but are focused on the crew's vital role in safety and mission success. Specifically, before every human spaceflight mission, FCOD is required to sign the CoFR, as documented in the Flight Crew Operations Space Flight Preparation Plan. As provided to the committee by NASA, the certification of crew readiness includes an assessment of the training and preparedness of the individuals and the entire crew. Addressing mission-specific readiness, the Mission Operations Directorate (MOD) signs a CoFR on completion of systems, EVA, robotics, payloads, in-flight maintenance, and emergency training. In addition, the FCOD certifies that both individual performance and combined crew performance are acceptable for the mission. The FCOD assessment is based on a compilation of training evaluations, instructor-astronaut evaluations (EVA, robotics, and emergency), peer evaluations for expeditionary training events, astronaut performance boards, T-38N skills and Crew Resource Management, and instructor-pilot evaluations. The training of NASA astronauts in aircraft is documented and tracked at the Aircraft Operations Division. Then, at the Flight Readiness Reviews, the heads of FCOD and MOD certify that the crew members are able and ready to conduct the mission safely and successfully.

CURRENT AND POST-SHUTTLE T-38N ASSETS

As the size of the Astronaut Corps has decreased, NASA has also reduced the number of T-38N aircraft that it operates and currently has less than half the number that it operated only 10 years ago (Figure 3.11). With the retirement of the space shuttle, the FCOD is planning to retire the four Shuttle Training Aircraft and to reduce the T-38N fleet from the current 21 to 16 by FY 2013. It is planned that as many as four T-38Ns will be placed in flyable storage at El Paso, Texas, for possible replacement and in the event that circumstances for fleet augmentation

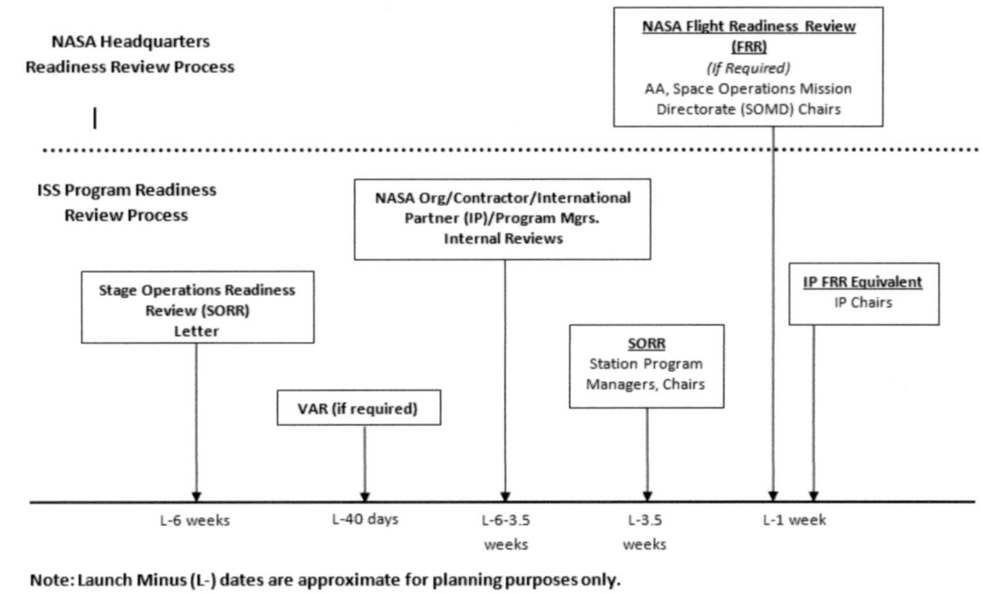

FIGURE 3.9 International Space Station Flight Readiness Review Process timeline. SOURCE: NASA Astronaut Office, "Ensuring the Readiness of the Astronaut Corps: A White Paper," March 25, 2011.

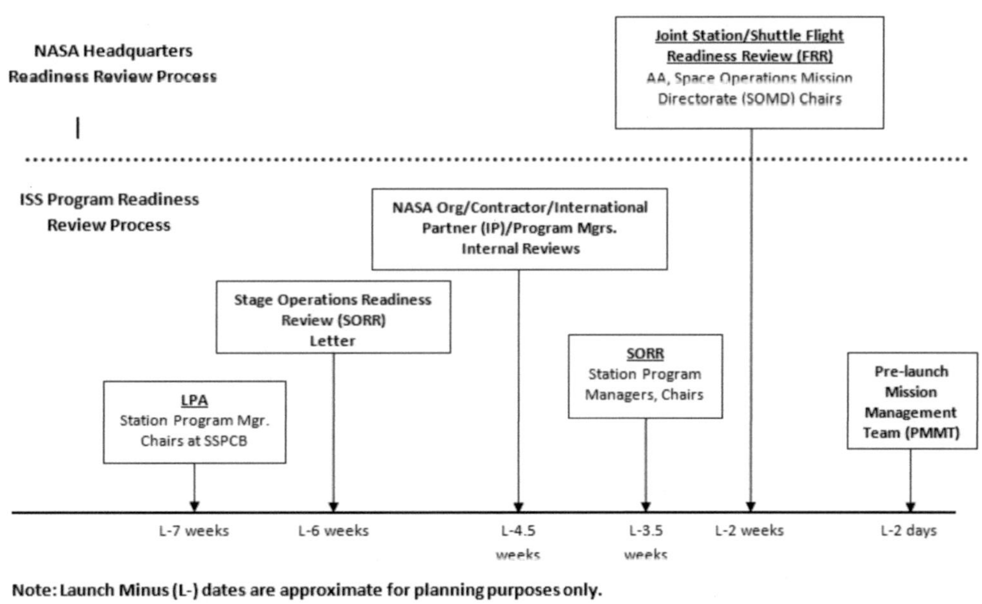

FIGURE 3.10 Soyuz Flight Readiness Review Process timeline. SOURCE: NASA Astronaut Office, "Ensuring the Readiness of the Astronaut Corps: A White Paper," March 25, 2011.

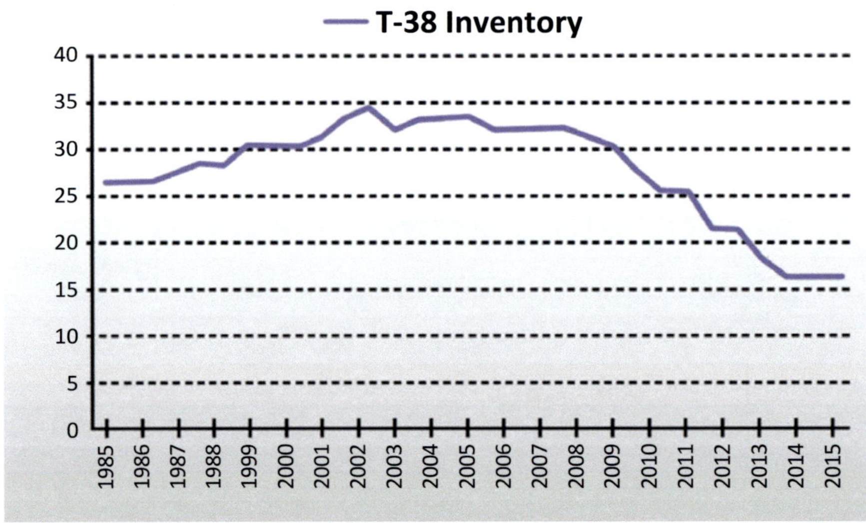

FIGURE 3.11 T-38N fleet size. SOURCE: NASA Johnson Space Center, Houston, Tex.

change (for example, demands from the commercial sector or other identified needs). This reduction will leave the FCOD with the T-38N and its reimbursable research aircraft: 2 WB-57s, 1 C-9, and 1 B377. The FCOD will also continue to operate a Gulfstream III mission support aircraft that will continue to be used for mission contingencies and to transport returning ISS crew members from the landing zone to Houston.[9]

The 2008 Fleet Planning Study

In March 2008, the FCOD culminated a multimonth Fleet Planning Study to determine the best aircraft or mix of aircraft to serve as its SFRT fleet. Several "attribute factors" were considered in this study. The aircraft or mix of aircraft must challenge experienced pilots, must train and develop inexperienced members of the Astronaut Corps, and must be able to instill spaceflight skills and attributes, specifically, discipline, priority-setting, crew coordination, communication, decision making, and spaceflight environment adaptation. As potential aircraft to improve SFRT, the aircraft evaluated in the Fleet Planning Study were the T-45 Goshawk U.S. Navy jet trainer aircraft, the T-6 Texan U.S. Air Force and Navy propeller trainer aircraft, the Premier 1A twin engine business jet, the Beech 400XP twin-engine business jet, the Learjet 60 twin-engine business jet (also in use with the FAA), the Cirrus single-engine propeller light aircraft, and the Lancair single-engine propeller light aircraft.

The primary conclusion of the Fleet Planning Study was that the T-38N provided more SFRT attributes than any other aircraft or combination studied and should be retained for continued use. In a fiscally unconstrained environment, the study asserted, having the option to supplement the T-38N fleet with another aircraft type might offer training advantages and, possibly, long-term cost savings, although maintaining proficiency in two aircraft would require further evaluation for both safety and astronaut availability. If a new aircraft fleet could be purchased, the Learjet 60's side-by-side seating, similar to Soyuz and CCS capsule configurations, could be valuable in developing crew teamwork and coordination skills applicable to the Soyuz. In addition, if the objective became allowing non-militarily trained astronauts to function as pilots in command, the T-6 would offer a pathway via an aircraft that, although highly capable, did not approach the performance of the T-38N. The study concluded that the T-38N was the best available alternative, especially inasmuch as there would be serious challenges in acquir-

[9] NASA Aircraft Operations Division, "Presentation to the NRC Committee on Human Spaceflight Crew Operations," presentation to the Committee on Human Spaceflight Crew Operations, January 6, National Research Council, Washington, D.C., 2011, pp. 6-7.

ing additional aircraft, including funding, competitive procurement requirements, and NASA Headquarters and congressional approval.

Although the committee lacked the time or resources to investigate the various aircraft options considered by NASA, it did review data concerning the possible acquisition of other high-performance two-seat jet aircraft from the military, such as the Air Force's F-16. They are newer than the T-38N but have a higher operating cost than the T-38N, apart from the cost of acquiring the aircraft and modifying them for NASA use.

POST-SHUTTLE SIMULATOR CAPABILITY

For astronauts aboard the ISS, essential spaceflight operation skills must be developed to meet the goal "to be able to operate as a team member in a highly dynamic, fast-changing, and sometimes unpredictable environment, with real-world, life-dependent consequences."[10] Crew members are required to learn crucial performance elements in (1) knowledge (what you know), (2) skills (what you can do), (3) rapid response (how you react), and (4) cognitive processing (how you think). Emergency response is incorporated into skills, rapid response, and cognitive processing.

The FCOD has made obvious and logical choices regarding which resources and facilities to retain and which to close. In a presentation before the committee, the MOD stated that its "task analysis process also identified minimum facility requirements necessary to complete training" and that its "requirements identified lowest cost minimum fidelity options for providing adequate crew training." Given that premise, the committee examined all training resources, including the T-38N fleet, which became its major focus. It is possible that some of the high-stress, time-critical training accomplished by SFRT and the T-38N could be performed, or at least augmented, in a high-fidelity ground-based ISS simulator; however, no such simulator exists, and its developmental cost may make this path prohibitive. Such a notional simulator would be capable of developing teamwork and rapid assessment of systems, conditions, and contingency planning in a time-constrained environment. Even with such characteristics, the committee concluded that this type of ISS simulator would still be unable to provide disorientation training and highly variable G-loading in connection with clear real-world consequences.[11]

No ISS simulator today can create a high-fidelity emergency training scenario. On-orbit training can compensate for that lack of fidelity, but it might also run the risk of creating a real emergency. Regardless of the current state of simulator capability, the committee notes that the ability to simulate emergencies on the ground would prove valuable, as NASA learned from the Apollo 13 experience.

DIFFERENCES BETWEEN SIMULATORS AND HIGH-PERFORMANCE FLIGHT ENVIRONMENTS

On the basis of its task to examine the SFRT program and its T-38N fleet, the committee formulated a straightforward question that could be posed by others both inside and outside government: Why is high-performance flight training required by NASA and DOD and not just ground-based simulator training before a first operational flight?

The short answer to the question is rooted in "risk" and "safety" or, put simply, it is to avert the loss of life, the loss of multibillion-dollar international assets, or mission failure. The expanded answer lies in the fact that there are external environmental factors in real flight that do not occur in simulators and that even if they could be simulated, they cannot be simulated without great expense—probably rivaling in-flight training in both capital investment and operating costs. Those factors affect operational performance physically and psychologically. If astronauts are not exposed to them in training before they experience them in real flight operations, both crew safety and mission success are exposed to risk and the possible loss of life.

When one examines the skill mix required of professional astronauts who operate spacecraft and their systems in nominal, off-nominal, and emergency situations, skills include academic achievement and technical knowledge,

---

[10] NASA Astronaut Office, "Presentation to the NRC Committee on Human Spaceflight Crew Operations," presentation to the Committee on Human Spaceflight Crew Operations, January 6, National Research Council, Washington, D.C., 2011, p. 61.

[11] NASA Astronaut Office, "Presentation to the NRC Committee on Human Spaceflight Crew Operations," presentation to the Committee on Human Spaceflight Crew Operations, January 6, National Research Council, Washington, D.C., 2011, pp. 61-64.

physical capabilities, and compatible psychological attributes. The academic and technical knowledge requirements for selection and mission training (for the specific vehicle and for unique mission attributes) are well documented. Astronaut selection criteria and general training curricula for the ISS, Soyuz, and international resupply vehicles have been examined as part of this study.

The physical and psychological attributes of astronaut candidates that are tested during selection are also well documented. However, manifestations of some of the attributes (such as performance during stressful situations) are not usually observed until astronauts are in real situations. A candidate may pass the psychological examination, but such an examination will not reveal whether he or she can work with another astronaut in an emergency procedure when both engines flame out at 41,000 feet over a thunderstorm or when power interruptions occur on orbit.

That line of candidate examination and committee questioning leads to exploring the principal differences between ground-based simulators and aircraft and spacecraft training. The differences are principally physical and psychological. For example, consider acceleration and G loads and the stress of the confined physical environment:

1. *Acceleration and G loads.* High-performance aircraft, such as the T-38N, accelerate from zero to Mach 1.3, exposing crewmembers to loads through their body and on their arms while they are controlling the vehicle, monitoring displays, and taking notes on their knee boards. The accelerations in the T-38N can expose crew members to up to +7.33 and down to −3.5 Gs. In comparison, the space shuttle topped out at 3 Gs, and on a Soyuz during ballistic re-entry, crew members experienced up to 8 Gs. In addition, the accelerations and the varying vector directions affect the vestibular system and can cause nausea. During all this, crew members must be focused on the vehicle and their individual crew position responsibilities. In the event of an emergency, they are required to focus on the task and ignore their environment. Ground-based simulators, even ones that are motion-based, do not expose crew members to accelerations and G loads, because emergency procedures are conducted in a normal 1 G environment.

2. *Confined stressful physical environments.* High-performance aircraft operating above 13,000 feet require that crew members wear helmets and oxygen masks even as they are restrained by harnesses in an ejection seat. Crew members are connected to supplemental oxygen, which forces oxygen into their masks in the event of a pressure loss. Similarly, crew members flying to the ISS, either in the past aboard the space shuttle or now and in the future aboard the Soyuz, launch in a pressure suit and helmet and strapped into a seat with a safety harness (NASA has currently contracted for six flights per year to be launched aboard Russia's Soyuz through 2016). Simulators are incapable of simulating the real claustrophobic environment of a spacecraft because a crew member can simply walk out of the confined area. In a spacecraft, that is impossible.

   a. Although the NASA space shuttle simulators could provide supplemental oxygen and suit cooling to simulate the in-flight suited environment, the experience clearly was missing the accelerations, unusual attitudes, and G loads. However, crew members were able to integrate all their "part task" training mentally because they had the knowledge of what those stressful environments felt like during their T-38N SFRT training.

   b. In comparison, the Soyuz trainer in Star City, Russia, allows the use of the Sokol pressure suit but is not motion based. The first exposure of the crew members to the actual vehicle is a few weeks before flight when they travel to Baikonur for a fit check in their custom-made Soyuz seat.

During the Apollo program, vehicle training stressors included heat and vibration. Crew members aboard the Soyuz also experience heat and vibration and may encounter them in future ballistic-type vehicles. Those stressors can still be experienced in high-performance jet training flights.[12]

3. *Exposure to a realistic psychological stress environment.* Astronauts training in high-performance jet aircraft are immersed in a fluid environment that demands constant mental attention, focus, and decision making. Unlike in a simulator, there is no "Pause" button in the aircraft to enable the crew members to escape challenging or deteriorating situations. Short of using their ejection seats, the crew must work together to stabilize and overcome

[12] NASA, *Flight Crew Operations Directorate Strategic Plan*, NASA Johnson Space Center, Houston, Tex., September 2006, p. 30.

whatever situation they encounter. Astronauts report that the high-stress aviation environment closely mimics the stresses of spaceflight and prepares them for the timely, accurate responses needed aboard a spacecraft.

EVOLVING TRAINING METHODS IN OTHER FIELDS

On the basis of the evidence presented to the committee, the Astronaut Office has established and maintains an effective training program to ensure mission safety and success. No other occupation has quite the same set of requirements, but some related industries have a number of the same issues. Notable among them is the operation of nuclear reactors by the Navy and by the civilian power generation industry. The operation of civil and commercial aircraft in the nation's air transportation system requires a number of the skills that are required by the astronauts. In those and other stressful, fast-paced occupations that require life-or-death decisions, research efforts are under way to examine and possibly improve the techniques and technology used to train individuals or teams. The substantial body of literature that is evolving will help to provide a more rigorous basis for many of the training methods in use and to develop training technologies and strategies to improve training outcomes.

In the commercial aviation industry, it is possible for a pilot to make his or her first landing of a revenue flight in a new aircraft type after only performing simulator training, that is, without conducting a training flight in the aircraft. However, that is not similar to first-time astronaut flight, for several reasons. The airline pilot who flies a new aircraft for the first time will already have many flight hours in other aircraft and probably will also have flown several flights in the new aircraft type under observation by an experienced captain. Although commercial aviation has increased the use of simulators, they are not adequate substitutes for flight experience.

The committee also heard from a representative of the Naval Nuclear Propulsion Program (typically referred to as Naval Reactors), which has adopted a training model that blends operational training in a fully functional reactor with supplemental training that uses a high-fidelity simulator. Its curriculum is focused on control room watch standers—a corollary with the ISS or any follow-on NASA vehicle. A basic underpinning of the Naval Reactors process is a coupling of the candidate's fundamental understanding of first principles and in-depth knowledge of system design and operation. In the Naval Reactors comparison, a level of competence must be demonstrated on watch in a fully functional reactor plant dedicated to candidate training, and watch standers are integrated into a team setting in which individual performance can be evaluated during both steady-state and simulated emergency situations. With that skill set as a foundation, Naval Reactors complements a person's training with simulated performance of routine and emergency tasks by using a computerized model of the propulsion plant that allows for skill development or remediation of deficient elements of performance. A foundation of the simulator training is maintaining an environment identical with that found in an operational power plant. The Naval Reactors approach is that a failure to do so could result in negative training, so deviations are scrupulously avoided. A further benefit of the Naval Reactors simulator model is that scenarios beyond the design basis can be examined to evaluate opportunities for improvement in design or operation. For spaceflight applications, it is reasonable to substitute a high-performance aircraft for the Naval Reactors training reactor and to supplement both curricula with computerized simulators to develop the candidate fully in a cost-efficient manner that blends safety and cost.

SUMMARY

NASA appears to have an effective plan for retiring shuttle-era ground-based facilities and is already implementing that plan. Most ISS training facilities will be retained. The committee's only disagreement with NASA concerns the retirement of the SES Dome. The committee concluded that there may be a near-term use for this facility and that NASA should study other possible uses for it.

Many required attributes cannot be trained for in simulators because they lack the spectrum of physical and psychological stresses encountered in spaceflight: accelerations, unusual attitudes, unexpected variables, pressure changes, external communications, and so on. There is also an inability in spaceflight to pause or freeze a situation or to walk away if a situation deteriorates to near loss of control.

Elimination of a high-performance airplane environment both for screening new "non-flight-experienced" astronaut crewmembers and for keeping all crew proficient in the attributes described above appears to introduce an unacceptable crew performance safety risk into the operation of multibillion-dollar spacecraft in return for relatively small savings in training costs.

That risk is not trivial and has been mitigated to date by a training and flight proficiency syllabus designed by both the Astronaut Office and professional instructor pilots—most of them with thousands of hours of prior military flight experience. Those subject-matter experts designed a syllabus that has demonstrated its success in (1) keeping trained pilots proficient in the skills for which they were selected as they wait for years between spaceflights, (2) bringing non-pilots to a skill level of safe spaceflight readiness and keeping them proficient, and (3) eliciting experience-based examples that show that successful performance in dire, real-life spacecraft situations was attributable in large part to the training experience gained in high-performance SFRT aircraft.

The value of SFRT is reinforced by the Ellington Field professional flight instructors, who have observed degradation in physical skills (specifically, hand-eye coordination) and decision-making skills when flight hours fall below the current syllabus levels—and the instructor pilots consider these currency hours and syllabus as providing a minimum level of training.

The end of the Space Shuttle program will substantially change many aspects of the U.S. human spaceflight program (Figure 3.12). It appears to the committee that high-performance aviation training is a high-confidence, experientially proven method for ensuring a common, minimum level of preparation for the dynamic, unpredictable, and hazardous environment of spaceflight that is certain to be encountered by U.S. astronauts in the coming decade and beyond. It is vital to protect and pass on the historical investment in the training and experience base that enables safe and successful human spaceflight operations. That legacy will help to ensure the success of future U.S. space exploration programs and commercial flight initiatives.

FIGURE 3.12 Space shuttle *Endeavour* at the International Space Station in May 2011. SOURCE: Courtesy of NASA.

FINDINGS AND RECOMMENDATIONS

Ground-Based Facilities

Finding 3.1. The NASA plan for post-shuttle retirement of shuttle-specific training facilities is generally appropriate. However, the Shuttle Engineering Simulator Dome may be useful in training for future activities, such as rendezvous and docking operations during commercial transportation of ISS crew.

Recommendation 3.1. NASA should evaluate potential future requirements for the Shuttle Engineering Simulator Dome and, if it will be needed, should preserve this facility.

Finding 3.2a. Now that the shuttle is retired, the specific spaceflight crew operations shift from shuttle operations and ISS assembly to Soyuz and ISS nominal and emergency operations, ISS payload operations, and ISS maintenance. The requirements for training of flight crews for those ISS operations include emergency response training, extravehicular activity operations, and the full suite of nominal operations for U.S. and international partner ISS elements, including Soyuz. Thus, the ISS ground-based training facilities are required for the support of crew training for future operations and maintenance of the ISS.

Finding 3.2b. The requirements for U.S. astronaut training include international partner ISS element operations at international partner facilities and Soyuz operations in Russia. The U.S. international partner agreements also require that the United States provide for enhancing skill proficiency and training for the international partner astronauts.

Recommendation 3.2. NASA should retain the capability and training facilities to conduct International Space Station (ISS) mission-specific training after retirement of the space shuttle to ensure the continued safety and mission success of ISS operations.

Spaceflight Readiness Training

Finding 3.3a. The spaceflight readiness training requirement is derived from safety and mission success requirements, not tied to any specific mission. Although the requirement is not expressly documented at the NASA Headquarters program level, it was developed by the Flight Crew Operations Directorate in response to NASA Headquarters-controlled safety and mission success requirements and embedded at the level of the NASA JSC Certificate of Flight Readiness for safe operations of flight, which is then provided to NASA Headquarters. Any changes in spaceflight readiness training need to be made with great care because changes can result in increased risk to safety and mission success.

Finding 3.3b. Spaceflight readiness training using high-performance aircraft has been demonstrated and documented to prepare crews for successful and safe spaceflight, dating back 50 years, from the inception of the Mercury program to the current International Space Station program. SFRT is more than just flying—the full spectrum of experiences gained is not restricted to the operation of high-performance aircraft but extrapolates to crew resource management and performance under stress. SFRT is used effectively internationally to produce qualified members of the Astronaut Corps who are independent of crew position or vehicle design.

Recommendation 3.3. To ensure continued safety and mission success, NASA should maintain a spaceflight readiness training program that includes high-performance aircraft.

Finding 3.4a. FCOD maintains the Astronaut Corps and provides the capability to conduct SFRT.

Finding 3.4b. High-performance aircraft present conditions, including crew disorientation and rapid fluctuation in G-forces, under which the flight crew must carry out complex tasks in a stressful and potentially life-threatening environment. That combination of unique environments, demand for rapid, critical decision making, and historical evidence convinced the committee that SFRT provides experience-based training that cannot be duplicated by current or, to the best of the committee's knowledge, projected alternative techniques or technologies.

Finding 3.4c. Given the current investment in the existing T-38N fleet, this fleet is the most cost-effective means of providing SFRT in the near term. In the long term, new technology that may be a more cost-effective means of providing SFRT might be demonstrated and proved.

Finding 3.4d. The size of the T-38N SFRT fleet is projected to fall to 16 aircraft in 2013.

Recommendation 3.4. NASA should retain the T-38N fleet for spaceflight readiness training and should fund the fleet at a level commensurate with the projected required size of the post-shuttle Astronaut Corps.

Learning from Other Occupations

Finding 3.5. Substantial research is being undertaken on selection and training of personnel in related high-stress occupations. Some of that work is leading to continually improving methods and technologies for training for team and individual performance in stressful high-risk situations.

Recommendation 3.5. NASA should continue to monitor training methods and technologies in related fields for possible ways to enhance the astronaut selection and training process.

Appendixes

A

Appropriate Training Methods and Technologies

A distinctive and challenging aspect of astronaut training is that astronauts must be fully functional crew members on their first flight, unlike the case in other safety-critical domains that have some measure of on-the-job training or which purposefully provide training by incrementally increasing duties within an operational environment. Thus astronaut training must be sufficient to meet all of the training requirements identified throughout this report.

The first training requirement is that astronauts be ready to perform specific, isolated tasks, and the second training requirement is that the depth of understanding required for each task be deliberately identified and established. The NASA Mission Operations Directorate currently has a task analysis methodology that is used to define tasks and to classify each task for each crew member according to whether the crew member will be a user, an operator, or a specialist; these classifications are then considered in the development of training facilities.

For the purposes of highlighting training methods, a more formal categorization is the Skills, Rules, Knowledge Framework offered by Rasmussen.[1]

> A knowledge-based behavior involves reasoning about the situation based on abstract knowledge of the situation and the available courses of action. This level of behavior is generally the first to be learned, and can be taught through training methods that focus on abstract reasoning, which includes classroom instruction and self-study. Thus, tasks which only need to be trained to this level do not require any unusual facilities or novel training methods.
>
> A rule-based behavior relates the immediate situation to rules and procedures. This type of behavior is sensitive to correctly recognizing the salient features of the immediate situation. Thus, training an astronaut to this type of behavior at a particular task requires experience in an environment sharing many of the relevant stimuli. For example, if astronauts are expected, after training, to be able to execute a robotics operation procedure, their training needs to be conducted in a facility that emulates the important dynamics of the operation, and simulates the important cues that correspond to each step in the procedure.
>
> A skill-based behavior is possible only with the greatest training, training in the most accurate emulations of the operational environment, and recent training. This type of behavior requires very little or no conscious control to perform or execute an action and is indicative of sensorimotor behaviors (in which the astronaut can smoothly sense relevant patterns in the environment and relate them to their required motor inputs) and of naturalistic decision making (also called recognition-primed or intuitive decision making). The most skilled behavior can be trained when the training facility closely emulates a wide range of aspects of the situation, but the skill will generalize to other tasks

[1] J. Rasmussen, Skills, rules, knowledge; signals, signs, and symbols, and other distinctions in human performance models, *IEEE Transactions on Systems, Man, and Cybernetics* 13:257-266, 1983.

requiring similar sensorimotor behaviors. Because these types of behavior allow for immediate and reliable responses, this depth of learning is particularly important to tasks resolving emergencies and to time-critical, safety-critical tasks.

Thus methods for training individual astronauts on specific tasks need to be tailored to the depth of knowledge expected for each task. For training to knowledge-based behaviors, classroom instruction and self-study are common. For rule-based behaviors, a mock-up or part-task simulation needs to emulate the key dynamics of the operation such that the astronaut can be expected to recognize the triggers of specific rules, identify the systems and controls to act on in order to execute the required steps, and then monitor the results to make sure that these steps are successful in terms of system response to their actions. Thus a mock-up or part-task simulation could be as simple as a computer simulation of the most important systems with similar controls, surrounded by a cardboard schematic of the surrounding environment.

Skill-based training requires the greatest-fidelity, highest-cost facilities. The NASA Neutral Buoyancy Laboratory, for example, is used in extravehicular activity (EVA) training; it situates the astronauts in a perceptibly risky environment so that they experience the discomfort and limited movement of a real space suit, and it strives to provide a realistic representation of International Space Station components such that the astronauts can quickly recognize important features from multiple perspectives.

Some industries, such as commercial aviation, have a significant population to train and have used an economy of scale to streamline training systematically. For example, airline pilots are first trained and tested on a range of knowledge-based behaviors in a ground school. Then, in training and testing on rule-based behaviors, the airline pilots move through a series of part-task simulators and cockpit mock-ups—these range from cardboard mock-ups of the entire cockpit (in which pilots are expected to learn the position of each cockpit control to the level of being able to reach each control with their eyes closed); to emulators of specific systems that they can run on their personal computers, imitating a system's operation by clicking the mouse on pictures of the correct buttons on the computer screen; to fairly complete mock-ups of the entire cockpit but without motion, sound, or the view out the window. Only when these knowledge- and rule-based behaviors are demonstrated do airline pilots move to the most advanced, highest-fidelity "Level-D" flight simulators. These simulators, according to Federal Aviation Administration regulations, must fit an extensive list of specific capabilities, including a full emulation of the cockpit in which all of the controls look and act exactly the same as those of the actual aircraft. These highest-fidelity simulators are associated with significant acquisition costs ($5 million to $20 million for established production systems) and operational costs (hundreds of dollars per hour), require specialized infrastructure (e.g., significant electrical power, reinforced concrete floors), and must be maintained by specialized personnel. The required use of these facilities is systematically justified through established, regulated methods for analyzing required tasks and their depth of understanding (such as the Advanced Qualification Program). Such training programs also monitor individual progression through the process in order to tailor training protocols to maximize both learning and cost-effectiveness, while routinely evaluating overall program efficacy and cost-effectiveness.

Many training environments, including those for the Astronaut Corps, do not have an economy of scale that warrants the acquisition and maintenance of a wide range of simulators of varying fidelity. Knowledge-based and rule-based behaviors can be learned in high-fidelity simulators, but skill-based behaviors cannot be learned in low-fidelity simulators or classrooms. Thus the need for the highest-fidelity training facilities is paramount, as smaller training operations must maintain high-fidelity training facilities and, for maximal cost-effectiveness, fully utilize them. Additionally, these facilities can eliminate any cheaper, lower-fidelity simulations, for which the trainees can instead train within the availability of higher-fidelity simulations.

A third training requirement is the need to develop teamwork skills in general and to execute these skills within a specific operational culture. These teamwork skills are largely assumed in astronaut training to emerge as a by-product of simultaneously training multiple individuals together (i.e., without formal teamwork training), although in other domains including commercial aviation, team training, such as crew resource management training, has become far more formalized than it is in astronaut training. Research suggests that team training interventions are a viable approach for enhancing team outcomes. Such training approaches are useful for improving cognitive outcomes, affective outcomes, teamwork processes, and performance outcomes. Moreover, results suggest that training content, team membership stability, and team size moderate the effectiveness of team training interventions. One function of this training is to establish a common operational culture that shares the same vernacular, reinforces team bonds, and establishes shared goals for performance and safety.

Finally, perhaps the least tangible aspect of training addresses the *meta-cognitive,* or executive, functions that astronauts must perform as part of their "cognitive control" while under stress. In their study of stress and human performance, Salas, Driskell, and Hughes define *stress* as the process by which certain environmental demands evoke an appraisal process in which perceived demand exceeds resources, and the result is undesirable psychological, physiological, or behavioral outcomes; for example, stress-related failures of decision making have been attributed to nearly half of fatal aviation accidents.[2] At a basic level, training for cognitive control and effective stress response is related to the development of executive functions that guide selective attention to appropriate aspects of the environment—such as pilots learning, when disoriented, to focus on a visual scan of flight instruments despite conflicting vestibular sensations. At a more holistic level, an expert is able to plan and pattern activities to avoid "cognitive lockup," to manage tasks effectively within given demands and resources, and to recognize effective decision-making strategies to apply to different types of situations.

Recent research suggests that three elements must be considered in training for effective cognitive control (including decision making) under stress. First, such training is most effective when the trainees enter actual operations with the perception that they are well prepared. Second, training theories suggest that the trainee should be trained on tasks and situations similar to those that will be experienced under stress, an effect referred to in the military as "train how you fight." Third, training for stressful tasks requires a stressful training environment.

The need for a stressful, operationally realistic training environment is also recognized operationally by safety-critical domains. For example, although commercial aviation often certifies its pilots on the basis of training in simulators alone, this industry also recognizes the need for a newly minted pilot then to fly real operations with a more experienced pilot for a significant portion of time as the new pilot develops further experience with the real operational environment.

Similarly, while military training increasingly uses simulators, live-fire exercises remain a vital component of training. For example, a review of the significant losses experienced by flight pilots in their first experiences of real combat motivated the ongoing U.S. Air Force Red Flag exercises, which re-create as realistically as possible the actual stresses of the real flight environment. Similarly, training for naval nuclear operations, for example, progresses from classroom and part-task simulator instruction through the training on actual nuclear power plants dedicated to training and providing the same hazards as those of real plants.

If one determines that a stressful training environment is necessary, in which environment should astronaut training be conducted? Historically, the operational culture of the Astronaut Corps has been centered on aviation, including its attention to safety and its valuing of teamwork and calm, systematic responses to emergencies. Bearing in mind that changing an operational culture is difficult and that the period of transition is a risk factor during which common safety nets within the organization are stressed, changing the Astronaut Corps training basis from aviation would be a significant, risky endeavor that should only be undertaken when there is a compelling reason.

Thus, the ideal training for the Astronaut Corps should be designed to integrate instructional content, instructional method, and training resources systematically and purposefully in a phased progression from classroom instruction, through simple procedural trainers (part-task simulators, mock-ups), through high-fidelity simulators, and ultimately into stressful training environments that foster an effective operational culture and which require response to stresses like those that may be experienced in spaceflight. For an Astronaut Corps using aviation as its shared operational experience, this ideal training environment would then include procedural trainers and simulators of spacecraft, and then, because the spacecraft are themselves unavailable for training, a transition to aviation environments that mirror the time pressure and physical stressors of spaceflight, including the discomforts of specialized suits, helmets, oxygen masks, and life-critical environmental support systems.

Even though ideal, such a full range of low- and high-fidelity simulators and aircraft as noted in the ultimate

[2] E. Salas, J.E. Driskell, and S. Hughes, Introduction: The study of stress and human performance, pp. 1-45 in *Stress and Human Performance* (E. Salas, and J.E. Driskell, eds.), Lawrence Erlbaum Associates, Inc., Mahwah, N.J., 1996.

training environment described above would be prohibitively expensive within current and foreseen budgets. The acquisition costs of specialized simulators, for example, would be significant. However, as noted throughout this appendix, the most important tasks must be trained to the level of skill-based behavior and to the extent that the astronauts can apply effective stress responses, and this requires training in real, stressful environments. Thus, sufficient training cannot be provided only in cheaper, low-fidelity simulations or classroom environments.

B

Acronyms and Glossary

Apollo	The 1961-1975 space program that followed the Gemini program.
ASCAN	Astronaut candidate.
ATV	Automated transfer vehicle. A European uncrewed resupply spacecraft.
C3PO	Commercial Crew and Cargo Program Office (NASA).
CCDev	Commercial Crew Development.
CoFR	Certification of Flight Readiness.
COTS	Commercial Orbital Transportation Services.
CRM	Crew resource management.
CSA	Canadian Space Agency.
CST-100	Space capsule in development by Boeing; it could carry up to seven astronauts to the International Space Station (ISS) and launch from a variety of rockets, including Atlas, Delta, and Falcon 9 rockets.
Cygnus	Space capsule in development by Orbital Sciences; it is designed to carry cargo from the Earth to the ISS and to dispose of ISS waste. It is not, however, designed for reentry into Earth's atmosphere and will not transport cargo back to Earth.
Dragon	Space capsule in development by Space Exploration Technologies (SpaceX); it is designed to carry cargo to and from the ISS.
Dream Chaser	Spacecraft in development by Sierra Nevada Corporation; it is designed to launch from an Atlas V launch vehicle and to carry up to seven astronauts and cargo to the ISS. It will return to Earth by way of a conventional runway landing, allowing it to bring back cargo from the ISS.
ESA	European Space Agency.
EVA	Extravehicular activity. Also known as a "spacewalk."
FCOD	Flight Crew Operations Directorate. Located at the NASA Johnson Space Center.
FRR	Flight Readiness Review.
Gemini	The 1965-1966 space program that was the follow-up to the Mercury program. Gemini missions included the first spacewalk.

GRT Generic robotics training.

HTV H-II Transfer Vehicle. A Japanese uncrewed resupply spacecraft.

ISS International Space Station.

JAXA Japan Aerospace Exploration Agency.
JSC Johnson Space Center.

LEO Low Earth orbit. Generally considered to be between 160 and 2,000 km above Earth's surface.

MCC Mission Control Center. Located at the NASA Johnson Space Center.
Mercury First human spaceflight program in the United States; it ran from 1959 to 1963 and included the first American to orbit Earth.
Mir Russian space station, operational from 1986 to 2001.
MMR Minimum Manifest Requirement. The minimum number of astronauts required to sustain a human spaceflight operation as determined by model input, including spaceflight program requirements and the 5-year rotation plan (see Chapter 2).
MOD Mission Operations Directorate. Located at the NASA Johnson Space Center.
MPCV Multi-Purpose Crew Vehicle (NASA). It is based on the design of the Orion Crew Exploration Vehicle.

NASA National Aeronautics and Space Administration.
NBL Neutral Buoyancy Laboratory. Located at NASA's Johnson Space Center, it is an astronaut training facility consisting of a large indoor pool of water used to simulate a low-gravity environment during astronaut training in the performance of various mission tasks.
NRC National Research Council.

Orion Orion Crew Exploration Vehicle. This spacecraft, in development by Lockheed Martin, is designed to carry four astronauts to the Moon or six astronauts to the ISS and to sustain a crew for 21.1 days.

PMMT Pre-launch Mission Management Team.

RMS Remote Manipulator System. Also known as the robotic arm used on the space shuttle.
RSA Russian Space Agency.

SES Dome Shuttle Engineering Simulator Dome.
SFRT Spaceflight readiness training. Operational environments in which the crew is trained for spaceflight.
Skylab Crewed U.S. space station in orbit from 1973 to 1979.
SLS Space Launch System.
SORR Stage Operations Readiness Review.
Soyuz Russian spacecraft used since the 1960s to deliver cosmonauts and astronauts to space.
Spacelab Reusable laboratory that flew on 22 space shuttle flights from November 1983 to August 1998.
SSPCB Space Studies Program Control Board.
Star City The site of the Gagarin Cosmonaut Training Center in Russia.

T-38N Two-person jet specifically outfitted for NASA's astronaut training needs.

UPT Undergraduate Pilot Training.

C

Committee and Staff Biographical Information

FREDERICK D. GREGORY, *Co-Chair,* is the managing director of aerospace and defense strategies at Lohfeld Consulting Group, Inc. He retired as the deputy administrator of NASA in 2005. Mr. Gregory's previous management positions at NASA include acting administrator, associate administrator for space flight, and associate administrator for the Office of Safety and Mission Assurance. He was selected to be an astronaut in 1978 and has logged 455 hours in space, including service piloting the space shuttle *Challenger,* on which he led a 7-day mission in 1985. He graduated from the United States Naval Test Pilot School and served as an engineering test pilot for the U.S. Air Force (USAF) and for NASA. When he retired from the Air Force as a colonel in 1993, he had logged approximately 7,000 hours of flying time in more than 50 types of aircraft. He has authored or co-authored several papers in the areas of aircraft handling qualities and cockpit design. He is a member or past member of numerous societies, including the Society of Experimental Test Pilots, the American Helicopter Society, the American Institute of Aeronautics and Astronautics (AIAA), and the Tuskegee Airmen. Mr. Gregory has been awarded the Defense Superior Service Medal and two Distinguished Flying Crosses and was designated an Ira Eaker Fellow by the Air Force Association. He holds a B.S. degree from the U.S. Air Force Academy and an M.S. degree in information systems from George Washington University.

JOSEPH H. ROTHENBERG, *Co-Chair,* is currently an independent consultant and senior vice president for international development at SSC (formerly Swedish Space Corporation). Previously he had been a senior adviser and president for the Universal Space Network, and served on its board of directors. He spent the last 4 years of his career at NASA Headquarters as associate administrator of space flight; in that position he was responsible for establishing policies and direction for the space shuttle and International Space Station (ISS) programs, as well as for space communications and expendable launch services. Prior to that, he served as director of NASA Goddard Space Flight Center, where he was responsible for space systems development and operations and for execution of the scientific research program for NASA Earth-orbiting science missions. He is widely recognized for leading the development and successful completion of the first servicing mission for the Hubble Space Telescope (HST), which corrected the telescope's flawed optics. Mr. Rothenberg holds an M.S. in engineering management from Long Island University. He has previously served on the NRC's Committee on Assessment of Options for Extending the Life of the Hubble Space Telescope, the Committee on Meeting the Workforce Needs for the National Vision for Space Exploration, the Beyond Einstein Program Assessment Committee, and the Steering Committee to Review Near-Earth Object Surveys and Hazard Mitigation Strategies.

MICHAEL J. CASSUTT is currently an adjunct professor at the University of Southern California. Mr. Cassutt is the co-author of two astronaut biographies—*Deke!* and *We Have Capture: Tom Stafford and the Space Race*—which address early issues in the formation of NASA's Astronaut Office. He is an experienced writer of nonfiction, not only contributing articles to such magazines as *Space Illustrated*, *Space World*, and *Air & Space,* but also to books such as *Magill's Survey of Science: Space Exploration Series.* He is also the author of the biographical encyclopedia *Who's Who in Space;* this work, for which Mr. Cassutt conducted dozens of interviews over a period of 10 years, contains biographies and photos of 700 astronauts and cosmonauts from around the world. Mr. Cassutt has appeared on camera for the History Channel and on BBC documentaries about disasters in space, as well as on specials about astronauts and test pilots. As an author, he has published more than three dozen short stories, a science fiction novel, and a fantasy novel and has co-edited an anthology. As a television script writer and producer, he has been on the staff of a dozen different primetime network and cable series. He received a B.A. in television and radio journalism from the University of Arizona.

RICHARD O. COVEY is an independent consultant. He retired in 2010 as president and chief executive officer (CEO) of United Space Alliance, LLC, where he was ultimately responsible for the direction, development, and operations of the company. United Space Alliance is NASA's prime contractor for space shuttle and ISS operations, including launch and recovery, mission planning and support, and astronaut training. Early in his career, Mr. Covey served in the USAF as a test force director, test pilot, and operational fighter pilot. As an operational fighter pilot, he flew 339 combat missions during two tours in Southeast Asia. Mr. Covey later joined NASA and became an astronaut, serving as pilot of Space Transportation System (STS) mission 51-I and STS-26, and as commander of STS-38 and STS-61. Mr. Covey also served as co-chair of the Return-to-Flight Task Group, for which he was awarded the NASA Distinguished Public Service Medal. His numerous honors also include the Department of Defense (DOD) Distinguished Service Medal, the DOD Superior Service Medal, 5 Air Force Distinguished Flying Crosses, 16 Air Medals, the Air Force Meritorious Service Medal, the Air Force Commendation Medal, the AIAA Haley Space Flight Award for 1988, and the American Astronautical Society Flight Achievement Award for 1988. He received a B.S. in engineering sciences with a major in astronautical engineering from the U.S. Air Force Academy and an M.S. in aeronautics and astronautics from Purdue University.

DUANE W. DEAL is currently senior vice president for National Security Programs at Stinger Ghaffarian Technologies, Inc. (SGT). Mr. Deal is a retired brigadier general in the USAF. Prior to joining SGT, he was the director of National Security Space Programs for the Johns Hopkins University Applied Physics Laboratory. Preceding his service at Johns Hopkins, Mr. Deal was commander of the Cheyenne Mountain Operations Center. His responsibilities in that role included executing the United States Northern Command homeland defense mission. Mr. Deal also commanded the Air Force's geographically largest wing, responsible for the nation's space surveillance and space control missions. He has extensive experience in USAF space operations, maintenance, and logistics, and also in flight operations where he piloted seven Air Force aircraft, including the SR-71. He has served on national commissions, including the Columbia Accident Investigation Board and the Defense Science Board Task Force. Mr. Deal received his B.S. in physics and an M.S. in counseling and psychology from Mississippi State University and an M.S. in systems management from the University of Southern California.

BONNIE J. DUNBAR is currently president and chief executive officer of Dunbar International, LLC. She is the former president and CEO of the Seattle Museum of Flight. Dr. Dunbar began her extensive career with NASA when she accepted a position as a payload officer/flight controller at NASA Johnson Space Center (JSC). She became a NASA astronaut and also served as deputy associate administrator for the Office of Life and Microgravity Sciences at NASA Headquarters. In 1994 and 1995, Dr. Dunbar lived in Star City, Russia, for 13 months to train as a backup crew member for a 3-month flight on the Russian space station Mir and was certified by the Russian Gagarin Cosmonaut Training Center to fly on long-duration Mir Space Station flights. In 1995 and 1996, she was detailed to the NASA JSC Mission Operations Directorate as assistant director; there she was responsible for chairing the International Space Station Training Readiness Reviews and facilitating Russian-U.S. operations and training strategies. Dr. Dunbar has also served as assistant director at the NASA JSC with a focus on university research, as deputy associate director for biological sciences and applications, and as associate director, technology integration and risk management. Dr. Dunbar retired from NASA in 2005. She received B.S. and M.S. degrees in ceramic engineering from the University of Washington and her Ph.D. in mechanical and biomedical engineering from the University of Houston. She is a member of the National Academy of Engineering.

WILLIAM W. HOOVER has held executive positions in national aviation, defense, and energy activities and in recent years has been a consultant in these areas. He is the former executive vice president of the Air Transport Association of America, representing the interests of the major U.S. airlines industry, particularly related to technical, safety, and security issues. Prior to holding this position, he served as the assistant secretary, defense programs, U.S. Department of Energy, where he was responsible for all aspects of the U.S. nuclear weapons development program. He is also a major general, U.S. Air Force (retired), and had responsible positions in the Air Force Space Program, within NATO, at the Pentagon with the Secretary of the Air Force, and in Vietnam, where he commanded a combat air wing and flew 97 missions as a fighter pilot. He has served as chair of the Aeronautics and Space Engineering Board of the National Research Council. He has been the chair or a member on behalf of the Academy on several studies for NASA, DOD, the Federal Aviation Administration (FAA), and the National Oceanic and Atmospheric Administration that resulted in significant policy or technology implementations. He has served on corporate boards and as a member of the NASA Advisory Council. He currently is on the board of the Virginia Air and Space Center, Hampton, Virginia. He holds a B.S. in engineering from the U.S. Naval Academy and an M.S. in aeronautical engineering from the Air Force Institute of Technology and is a Distinguished Graduate of the National War College and a lifetime National Associate of the National Academies.

THOMAS D. JONES is senior research scientist with the Florida Institute for Human and Machine Cognition and a planetary science consultant to NASA and the aerospace community. He is a writer and speaker, and he serves on the board of directors of the Association of Space Explorers. As a NASA astronaut and mission specialist, Dr. Jones logged more than 52 days in space on four shuttle missions. He helped direct science operations on STS-59 (Space Radar Laboratory 1), was payload commander for STS-68 (Space Radar Laboratory 2), and helped deliver the Destiny laboratory to the ISS on STS-98. His previous positions include senior scientist for the Science Applications International Corporation, program management engineer at the Central Intelligence Agency, and B-52 pilot and aircraft commander for the USAF. He has published six books and is a member of the American Astronomical Society and the American Geophysical Union. Dr. Jones received a B.S. in basic sciences from the U.S. Air Force Academy and a planetary science Ph.D. from the University of Arizona.

FRANKLIN D. MARTIN is the president of Martin Consulting, Inc., which provides services in aerospace, including his participation on review boards for NASA flight projects. Dr. Martin has also been working with 4-D Systems since 2002. Sponsored by NASA's Office of the Chief Engineer's Academy of Program/Project and Engineering Leadership, the major focus of 4-D Systems is performance enhancement for NASA teams. His career with NASA and Lockheed Martin includes the following: science mission operations on Apollo 16 and Apollo 17; director, Solar Terrestrial and Astrophysics at NASA Headquarters (included the Sounding Rocket and Balloon Programs); Goddard Space Flight Center director for Space and Earth Science; NASA deputy associate administrator, Space Station; NASA associate administrator for Human Exploration; and director, Space Systems and Engineering, Civil Space for Lockheed Martin, with responsibility for the HST servicing missions, Space Infrared Telescope Facility (Spitzer), Lunar Prospector, and Gravity Probe-B. Dr. Martin resigned from NASA in 1990 and retired from Lockheed Martin in 2001. He received a B.A. with majors in physics and in mathematics from Pfeiffer University and a Ph.D. in physics from the University of Tennessee.

HENRY McDONALD is the Distinguished Professor and Chair of Computational Engineering at the University of Tennessee at Chattanooga. Dr. McDonald worked in the aerospace industry of the United Kingdom on a number of civil and military aircraft before immigrating to the United States. In the United States, he was a staff member at United Technologies Research Center, where he concentrated on turbomachinery, which eventually became known as computational fluid dynamics. Dr. McDonald then formed Scientific Research Associates, a small research and development company. While at Scientific Research Associates, he was asked to assist the NASA team investigating the *Challenger* disaster. Subsequently he became a member of the Lockheed Martin team investigating a Titan motor failure. Dr. McDonald later held a number of academic posts at Pennsylvania State University and Mississippi State University before becoming director of NASA Ames Research Laboratory, a position that he held from 1996 to 2002. Dr. McDonald is a member of the National Academy of Engineering, a fellow of the Royal Academy of Engineering, a fellow and honorary member of the American Society of Mechanical Engineers, an honorary fellow of the AIAA, and a fellow of the Royal Aeronautical Society.

AMY R. PRITCHETT is currently the David S. Lewis Associate Professor of Cognitive Engineering in the School of Aerospace Engineering of the Georgia Institute of Technology, holding a joint appointment in the School of Industrial and Systems Engineering. Dr. Pritchett has led numerous research projects sponsored by industry, NASA, and the FAA. Via the Intergovernmental Personnel Act, she served as director of NASA's Aviation Safety Program; she was responsible for the planning and execution of the program, conducted at four NASA research centers, where she sponsored roughly 200 research agreements and served on several executive committees, including the Office of Science and Technology Policy's Aeronautic Science and Technology Subcommittee and the executive committees of the Commercial Aviation Safety Team (CAST) and the Aviation Safety Information Analysis and Sharing program. Dr. Pritchett has had more than 170 scholarly publications in conference proceedings and in scholarly journals such as *Human Factors, Journal of Aircraft,* and *Air Traffic Control Quarterly*. She received the RTCA, Inc., William H. Jackson Award and, as part of CAST, the Collier Trophy. She is a member of the FAA Research, Engineering, and Development Advisory Committee (REDAC) and chairs the Human Factors REDAC subcommittee. She is a licensed pilot of airplanes and sailplanes. Dr. Pritchett received S.B., S.M., and Sc.D. degrees in aeronautics and astronautics from the Massachusetts Institute of Technology.

RICHARD N. RICHARDS is an independent consultant. He retired from the Boeing Corporation in 2007 as the deputy program manager for the space shuttle. At Boeing he also worked as director of New Reusable Systems and as program director for Shuttle and Space Station Integration. Previously, as an astronaut for NASA, Mr. Richards piloted and was commander of STS-28, STS-41, STS-50, and STS-64; he logged almost 34 days in space. While at NASA he also served as mission director for the Space Shuttle program office and as manager for the Space Shuttle program integration. Before his service at NASA, Mr. Richards had been a test pilot and naval aviator for the U.S. Navy, retiring after 25 years of service. He flew various aircraft in active squadrons in the U.S. Navy. He has been awarded the Distinguished Flying Cross, the NASA Distinguished Service Medal, the National Defense Service Medal, and the Vietnam Service Medal, among other awards. Mr. Richards received a B.S. in chemical engineering from the University of Missouri and an M.S. in aeronautical systems from the University of West Florida.

JAMES D. VON SUSKIL is currently the vice president of nuclear oversight at NRG Texas, representing that company's interests as part-owner of the South Texas Project nuclear power plant. Mr. von Suskil retired from the nuclear submarine force in 1995, having commanded the attack submarine USS *Augusta* and then a submarine squadron, and having served as chief of staff of the Pacific Submarine Force. After his retirement from the U.S. Navy, he worked for PECO Energy and was vice president of the Limerick Generation Station when PECO merged with Commonwealth Edison to form Exelon Corporation. For the next 3 years he served as vice president of Braidwood Generating Station before he left Exelon to become an independent consultant. Mr. von Suskil has served on various safety review boards and as a technical problem solver, including his work as special assistant for training in Naval Reactors, consultant to NASA on the Project Prometheus nuclear reactor program, and membership on the Atlantic Fleet Nuclear Propulsion Examining Board and the Nuclear Fuel Services Safety Culture Board of Advisors. After performing due-diligence services for NRG Energy in anticipation of its purchasing the South Texas Project, he became a full-time employee and currently works from the company's Houston office. He graduated from the U.S. Naval Academy and earned an M.S. in mechanical engineering from the Naval Postgraduate School.

Staff

DWAYNE A. DAY, *Study Director,* is a senior program officer for the National Research Council's (NRC's) Aeronautics and Space Engineering Board (ASEB). He has a Ph.D. in political science from the George Washington University. Dr. Day joined the NRC as a program officer for the Space Studies Board (SSB). Before that he had served as an investigator for the Columbia Accident Investigation Board, was on the staff of the Congressional Budget Office, and also worked for the Space Policy Institute at the George Washington University. He has held Guggenheim and Verville fellowships and was an associate editor of the German spaceflight magazine *Raumfahrt Concrete,* in addition to writing for such publications as *Novosti Kosmonavtiki* (Russia), *Spaceflight,* and *Space Chronicle* (United Kingdom). He has served as study director for several NRC reports, including *Space Radiation Hazards and the Vision for Space Exploration* (2006), *Grading NASA's Solar System Exploration Program: A Midterm Review* (2008), and *Opening New Frontiers in Space: Choices for the Next New Frontiers Announcement of Opportunity* (2008).

CATHERINE A. GRUBER, editor, joined the SSB as a senior program assistant in 1995. Ms. Gruber first came to the NRC in 1988 as a senior secretary for the Computer Science and Telecommunications Board and also worked as an outreach assistant for the National Science Resources Center. She was a research assistant (chemist) in the National Institute of Mental Health's Laboratory of Cell Biology for 2 years. She has a B.A. in natural science from St. Mary's College of Maryland.

LEWIS GROSWALD, a research associate, joined the SSB as the Autumn 2008 Lloyd V. Berkner Space Policy Intern. Mr. Groswald is a graduate of George Washington University, where he received a master's degree in international science and technology policy and a bachelor's degree in international affairs, with a double concentration in conflict and security and Europe and Eurasia. Following his work with the National Space Society during his senior year as an undergraduate, Mr. Groswald decided to pursue a career in space policy, with a focus on educating the public on space issues and formulating policy.

AMANDA R. THIBAULT, a research associate, joined the ASEB in 2011. Ms. Thibault is a graduate of Creighton University, where she earned her B.S. in atmospheric science in 2008. From there she went on to Texas Tech University where she studied lightning trends in tornadic and non-tornadic supercell thunderstorms and worked as a teaching and research assistant. She participated in the VORTEX 2 field project from 2009 to 2010 and graduated with an M.S. in atmospheric science from Texas Tech in August 2010. She is a member of the American Meteorological Society.

DIONNA WILLIAMS is a program associate with the SSB, having previously worked for the NRC's Division of Behavioral and Social Sciences and Education for 5 years. Ms. Williams has a long career in office administration, having worked as a supervisor in a number of capacities and fields. She attended the University of Colorado at Colorado Springs and majored in psychology.

MICHAEL H. MOLONEY is the director of the SSB and the ASEB at the NRC. Since joining the NRC in 2001, Dr. Moloney has served as a study director at the National Materials Advisory Board, the Board on Physics and Astronomy (BPA), the Board on Manufacturing and Engineering Design, and the Center for Economic, Governance, and International Studies. Before joining the SSB and ASEB in April 2010, he was associate director of the BPA and study director for the Astro2010 decadal survey for astronomy and astrophysics. In addition to his professional experience at the NRC, Dr. Moloney has more than 7 years' experience as a foreign-service officer for the Irish government; he served in that capacity at the Embassy of Ireland in Washington, D.C., the Mission of Ireland to the United Nations in New York, and the Department of Foreign Affairs in Dublin, Ireland. A physicist, Dr. Moloney did his graduate Ph.D. work at Trinity College Dublin in Ireland. He received his undergraduate degree in experimental physics at University College Dublin, where he was awarded the Nevin Medal for Physics.